34th Annual North American Membrane Society Annual Meeting (NAMS 2025)

Emerging Frontiers

Nashville, Tennessee, USA
17-21 May 2025

ISBN: 979-8-3313-2773-6

Printed from e-media with permission by:

Curran Associates, Inc.
57 Morehouse Lane
Red Hook, NY 12571

CURRAN ASSOCIATES INC. proceedings.com

Some format issues inherent in the e-media version may also appear in this print version.

North American Membrane Society (NAMS)
University of Arkansas
Ralph E. Martin Department of Chemical Engineering
3202 Bell Engineering Center
Fayetteville, AR 72701-1201
USA

Phone: +1 479 575-7926
Fax: +1 479 575-7926

names@uark.edu

Additional copies of this publication are available from:

Curran Associates, Inc.
57 Morehouse Lane
Red Hook, NY 12571 USA
Phone: 845-758-0400
Fax: 845-758-2633
Email: curran@proceedings.com
Web: www.proceedings.com

North American Membrane Society

34th Annual Meeting

Emerging Frontiers

May 17-21th
2025

Conference Chairs

Xiaobo Dong, Ph. D.
Alexander Lopez, Ph. D.
Piran Kidambi, Ph. D.

www.membranes.org/nams-2025

Table of Contents

Dear NAMS Community,

On behalf of the conference organizers, it is our great pleasure to welcome you to the 2025 NAMS Annual Meeting at the Loews Hotel in Nashville, Tennessee. Whether this is your first time attending or you're a returning participant, we're excited to have you join us.

This year's meeting, themed **"Emerging Frontiers",** aims to explore new horizons in membrane science and technology. With an outstanding lineup of technical sessions, keynote addresses from some of the most influential voices in the field, and engaging workshops led by thought leaders, the program is designed to spark meaningful conversations and drive innovation across our community.

A highlight of the week will be Monday evening's poster session, featuring cutting-edge research from across the community. This session not only showcases exciting new developments but also provides a platform to support and engage with the next generation of membrane scientists through the student poster competition. We encourage all attendees to explore the posters, speak with student presenters, and learn about their work, aspirations, and passion for science.

To our first-time attendees and students, a special welcome! We hope you find this conference to be a welcoming space to discover the field, build connections, and begin forging your path within our vibrant and supportive community. To our long-time attendees, volunteers, exhibitors, and sponsors, thank you. Your continued support makes this event possible and helps propel the field forward.

We invite everyone to make the most of this experience: ask questions in person or virtually on Whova, share ideas, meet new colleagues, and reconnect with old friends. And while you're here, take time to enjoy all that Nashville has to offer—from its legendary live music to its world-class cuisine and warm Southern hospitality.

Thank you for being part of this year's conference. We look forward to an exciting and productive week together.

Warm regards,

Xiaobo Dong, Alex Lopez, Piran Kidambi
The 2025 NAMS Annual Meeting Co-chairs

NAMS 2025 Anti-Harassment Policy

NAMS 2025 is dedicated to providing a harassment-free conference experience for everyone. We will not tolerate harassment of conference participants (including sexual harassment and assault) in any form. Conference participants violating these rules may be sanctioned or expelled from the conference without a refund at the discretion of the conference organizers. Harassment includes but is not limited to: comments related to gender, gender identity, and expression, sexual orientation, physical appearance, race, or religion; unwanted sexual attention; unwanted physical contact; threats of violence; intimidation; and stalking.

Reporting: If someone makes you or anyone else feel unsafe or unwelcome, please report it as soon as possible to one of the conference co-chairs. If you are being harassed by someone, notice that someone else is being harassed, or have any other concerns, please contact one the conference co-chairs through the Whova App or in person.

Enforcement: If a participant engages in harassing behavior, event organizers retain the right to take any actions to keep the event a welcoming environment for all participants. This includes warning the offender or expulsion from the conference with no refund. Event organizers may take action to redress anything designed to, or with the clear impact of, disrupting the event or creating a hostile environment for attendees. We expect participants to follow these rules at all event venues and event-related social activities.

Welcome to NAMS 2025 in Nashville

Fellow Members & Attendees:

Welcome to another stimulating installment of the North American Membrane Society annual meeting. The latest numbers show an above average number of registrants which exemplifies the excitement in membrane science and commitment from our community.

The organizers have built a dynamic program that aims to represent the breadth of discoveries and scientific advances in our field. There will be **SIX** parallel sessions this year through each day. We have also added **TWO** new workshops to represent the growth in membrane science: "*Electrochemical Process Using Membranes*" and "*Introduction to Modeling and Scale-up Concepts*". These workshops, along with the classic topics, are taught by a cast of experts from academia, industry, and national labs. And as always, we have a plenary speaker lineup comprised of membrane VIPs that will keep the mental gears turning.

Of course, it's not all work and no play. The opening reception should prove to be an exciting adventure for the group as we explore cultural highlights in Nashville on Sunday. And we culminate with a grand banquet on the final evening to celebrate and recognize our members for their achievements.

Don't forget to stop by and see the latest technology offerings from our numerous exhibitors. Also, please thank a sponsor if you see them. Their generous contributions provide support for our students to participate in this premier event.

Looking forward to seeing everyone in Nashville. You probably won't see me in the honky tonks on Broadway…*probably*?

Cheers,

JR Johnson

NAMS President

president@membranes.org

Event Sponsors

gold level

MilliporeSigma

silver level

the human energy company® Chevron

OSMOSES

AIR PRODUCTS

Membrane Solutions

AMERIDIA INNOVATIVE SOLUTIONS

novo nordisk®

DOW®

Bronze level

Gale and Dibakar Bhattacharyya, Ph. D.

Friend of NAMS

Georgia Tech ISSEC

PROMETHEUS
CARBON NEUTRAL E-FUELS

Exhibitor

APTCO TECHNOLOGIES

VERICO
TECHNOLOGY

porometer

Surface Measurement Systems
World Leader in Sorption Science

icspi

3DMem
Print Your Membrane

SGMEM
Singapore Membrane Consortium

2025 NAMS Fellows

Suzana Pereira Nunes

KAUST Vice Provost for Faculty and Academic Affairs

Professor, Chemical and Environmental Science and Engineering

King Abdullah University of Science and Technology (KAUST)

Jamie A. Hestekin

Maurice E. Barker Chair in Chemical Engineering

Ralph E. Martin Department of Chemical Engineering

University of Arkansas

William Bernard Krantz

President's Teaching Scholar & Professor Emeritus

Department of Chemical and Biological Engineering

University of Colorado

Elias Klein Travel Supplement recipients

Supports students needing limited financial assistance. This award program is named in honor of Elias Klein, whose vision and spirit guided the founding of NAMS in 1985.

Name	Institution
Adbulkadir Muhammad Zabi	Clarkson University
Akanksha Majumder	University of Connecticut
Anelisse Claros	University of Arkansas
Azmain Akash	Louisiana State University
Bar Avneri	Technion-Israel Institute of Technology
Chenhao Yao	The University of Texas at Austin
Claire Velikonja	McMaster University
Deepanjali Roy	West Virginia University
Himangshu Mondal	The University of Alabama
Jacalyn Morgan	University of Colorado Boulder
Jocelyn A Riet	University of Toronto
Josh Goodrich	University of Arkansas
Juliana Marcal	University of Arkansas
Kritika Jashrapuria	Indian Institute of Technology Bombay
Md Abdullah Al Masud	West Virginia University
Muhammad Usman Yousaf	University of Kentucky
Nikan Mokarinezhad	Tufts University
Toheeb Olumide Obidara	Northeastern University
Vaishnavi Koyilath Nandakumar	Columbia University
Yongha Kim	Pennsylvania State University

NAMS
The North American Membrane Society

Thank you reviewers for the Student Fellowship & Young Membrane Scientist awards!

Reviewer	Institution
Albert Kim	University of Hawaii
Ayse Asatekin	Tuffs University
Christina Carbrello	MilliporeSigma
Christine Duval	Case Western Reserve University
David Hopkinson	National Energy Technology Laboratory
Emily Rabe	Memberion
Hee Jeung Oh	Pennsylvania State University
Jason Williams	Shell
John Pellegrino	University of Colorado, Boulder
JR Johnson	Avnos
Mahdi Malmali	Texas Tech University
Margaret Lee	Pall
Marie-Eve Langevin	Ameridia
Melinda L. Jue	Lawrence Livermore National Laboratory
Priyesh Wagh	Lubrizol
Ranil Wickramasinghe	University of Arkansas
Syed Islam	Oakridge National Laboratory
Uwe Beuscher	W.L. Gore & Associates
William Phillip	University of Notre Dame
Yifu Ding	University of Colorado, Boulder

NAMS
The North American Membrane Society

Student Fellowship Award

Outstanding graduate students in the membrane science and technology area

Ronald (RJ) Vogler

The University of Texas at Austin

David Lu

University of Kentucky

Sean M. Bannon

The University of Virginia

Vinh Bui

University at Buffalo

NAMS
The North American Membrane Society

Young Membrane Scientist

Outstanding individuals who are starting their professional careers in membrane science and technology
(≤5 yr from terminal degree)

Dr. Thien (James) Tran
National Energy Technology Laboratory

Dr. Mahmoud Atif Abdulhamid
King Fahd University of Petroleum and Minerals

Dr. Taehoon Lee
Sungkyunkwan University

NAMS
The North American Membrane Society

Honorable mentions

Young Membrane Scientist & Student Fellowship Award

Dr. Mengyang Hu
Kobe University

Agboola Suleiman
University of Notre Dame

Lauren M. Ward
The University of Alabama

NAMS
The North American Membrane Society

Permeance Prize for Mid-career Excellence

Outstanding individual who are in the middle of their professional careers in membrane science & technology
(≤15 yr from terminal degree)

Prof. Ayse Asatekin
Tufts University

Prof. Manish Kumar
The University of Texas at Austin

NAMS
The North American Membrane Society

2025 North American Membrane Society Annual Meeting

May 17th – 21st, 2025

Loews Vanderbilt Hotel

Nashville, TN

Conference Agenda

Saturday May 17th, 2025

Time	Event	Location
8:00 AM – 5:00 PM	Conference Registration	Starstruck Gallery
8:00 AM – 5:00 PM	Workshop 1 - Water Treatment	Blackbird Studio AB
8:00 AM – 5:00 PM	Workshop 2 - Measurement	Southern Ground AB
8:00 AM – 5:00 PM	Workshop 3 - Electrochemical	Platinum Room

Sunday May 18th, 2025

Time	Event	Location
8:00 AM – 6:00 PM	Conference Registration	Starstruck Gallery
8:00 AM – 5:00 PM	Workshop 4 - Gas Separations	Blackbird Studio AB
8:00 AM - 5:00 PM	Workshop 5 - Modeling	Platinum Room
8:00 AM – 5:00 PM	NAMS Board Meeting	Green Room
2:00 PM – 5:00 PM	Student Workshop	Southern Ground AB
6:00 PM – 9:00 PM	***Welcome Reception***	***National Museum of African American Music***

Monday May 19th, 2025

Time	Event	Location
8:00 AM – 6:00 PM	Conference Registration	RCA
8:00 AM – 10:30 AM (Parallel Session A) (5 talks each) (30 total)	Water Treatment I	Blackbird Studio A
	Ion/Ion Separations I	Blackbird Studio B
	Polymer Design/Synthesis I	Southern Ground AB
	NAMS Awards Session I	Ocean Way
	Virus Filtration	Green Room
	Carbon Capture and Sequestration I	Sound Emporium A
10:30 AM – 11:00 AM	Coffee Break	Starstruck Gallery
11:00 AM – 12:00 PM	***Plenary Lecture 1 – Bill Koros***	***Symphony Ballroom***
12:00 PM – 2:00 PM	***Lunch with Legends***	
2:00 PM – 4:00 PM (Parallel Session B) (4 talks each) (24 total)	Water Treatment II	Blackbird Studio A
	Ion/Ion Separations II	Blackbird Studio B
	Polymer Design/Synthesis II	Southern Ground AB
	NAMS Awards Session II	Ocean Way
	Modeling Gas/Vapor Transport	Green Room
	Carbon Capture and Sequestration II	Sound Emporium A
4:00 – 4:30 PM	Coffee Break	Starstruck Gallery
4:30 PM – 6:30 PM (Parallel Session C) (4 talks each) (24 total)	Hollow Fiber and Gas Membrane Design	Blackbird Studio A
	Dehydration/Dehumidification Separations	Blackbird Studio B
	Hydrogen Separations	Southern Ground AB
	Membrane Fouling/Biofouling I	Ocean Way
	High Salinity Streams and Unconventional Systems	Green Room
	Biosystems and Bioreactors I	Sound Emporium A
7:30 PM – 10:00 PM	***Poster Session***	**Symphony Ballroom**

Tuesday May 20th, 2025

Time	Event	Location
8:00 AM – 12:00 PM	Conference Registration	RCA
8:00 AM – 10:30 AM (Parallel Session D) (5 talks each) (30 total)	Ion Transport Investigations I	Blackbird Studio A
	WAMS Session	Blackbird Studio B
	Medical Applications	Southern Ground AB
	Critical Mineral and Resource Recovery	Ocean Way
	Aqueous Phase Modeling	Green Room
	Nanofiltration and Ultrafiltration I	Sound Emporium A
10:30 AM – 11:00 AM	Coffee Break	Starstruck Gallery
11:00 AM – 12:00 PM	***Plenary Lecture 2 – Beate MT***	***Symphony Ballroom***
12:00 PM – 1:30 PM	Lunch Break	
1:30 PM – 3:30 PM (Parallel Session E) (4 talks each) (24 total)	Ion Transport Investigations II	Blackbird Studio A
	Mixed Matrix Membranes	Blackbird Studio B
	Molecular Simulations	Southern Ground AB
	Membrane Fouling/Biofouling II	Ocean Way
	Biofilms and Biosystems Recovery	Green Room
	Nanofiltration and Ultrafiltration II	Sound Emporium A
3:30 – 4:00 PM	Coffee Break	Starstruck Gallery
4:00 PM – 6:00 PM (Parallel Session F) (4 talks each) (24 total)	Ion Transport Investigations III	Blackbird Studio A
	Membrane Distillation	Blackbird Studio B
	Electrochemical Separations I	Southern Ground AB
	Olefin/Paraffin Separations	Ocean Way
	Organic Solvent Separations	Green Room
	Biosystems and Bioreactors II	Sound Emporium A
6:00 PM – 7:00 PM	NAMS Business Meeting	Ocean Way
7:00 PM – 7:30 PM	***Cocktail Hour***	***Symphony Foyer***
7:30 PM – 10:00 PM	***NAMS Banquet and Awards***	***Symphony Ballroom***

Wednesday May 21st, 2025

Time	Event	Location
8:00 AM – 10:30 AM (Parallel Session G) (5 talks each) (30 total)	Biobased Separations	Blackbird Studio A
	Composite Membranes	Blackbird Studio B
	Biomimetic and Bioinspired Membranes	Southern Ground AB
	Desalination and Osmotic Processes I	Ocean Way
	Material and Module Design and Aging	Melody Room
	Molecular Sieve Separations	Sound Emporium A
10:30 AM – 11:00 AM	Coffee Break	Starstruck Gallery
11:00 AM – 12:00 PM	***Plenary Lecture 3 - Winston Ho***	***Symphony Ballroom***
12:00 PM – 1:45 PM	Lunch Break	
1:30 PM – 3:30 PM (Parallel Session H) (4 talks each) (20 total)	Circular Economy	Blackbird Studio A
	Module and Systems Modeling	Blackbird Studio B
	Electrochemical Separations II	Southern Ground AB
	Desalination and Osmotic Processes II	Ocean Way
	Inorganic Membranes	Melody Room

Technical Sessions Summary

5 Workshops | 3 Plenary Lectures | 47 Sessions with 206 30-minute Oral Presentations

Monday	Water Treatment I	Ion/Ion Separations I	Polymer Design/Synthesis I	NAMS Awards Session I	Virus Filtration	Carbon Capture and Sequestration I
Location	Blackbird Studio A	Blackbird Studio B	Southern Ground AB	Ocean Way	Green Room	Sound Emporium A
Session chairs	Steven Weinman	Hani Hawa	Jason Williams	Xiaobo Dong	Cristiana Boi	David Hopkinson
	Laura H. Arias Chavez	Joshua Moon	Margaret Lee	Piran Kidambi	Gosia Chwatko	Jason Bara
8:00-8:30	369- Characterization, Performance Evaluation, and Toxicological assessment of Polysulfone-Sulfonated Polyether Ether Ketone Membranes for Water separation Applications Author: Muhammad Usman Yousaf	47- Poly(Terphenyl Fluorenyl Isatin) with Fluorene-Linked Sulfonate Groups and Microblock Structures For Proton Exchange Membranes Author: Chaoyi Ba	64- Structure-Property-Performance Relationships of Upcycling Waste PMMA in PVDF Membrane Design Author: Joseph Danner	344- Self-assembling, charge-containing polymers for highly selective and fouling-resistant membranes Author: Ayse Asatekin	108- Investigating the electric field driven membrane separation of biomolecules towards increased viral vector purity Author: Emily Ingram	393- Bringing MTR's Polaris™ membrane CO2 capture process to commercial scale Author: Witopo Salim
8:30-9:00	58- Design principles of reactive nanofiltration membranes for water treatment Author: Yanghua Duan	360- Looking beyond size and charge: A roadmap to ion separations for the global energy transition Author: Ryan Kingsbury	104- Adhesion behavior in micro/ultra-filtration membranes bonded under industry-relevant conditions Author: Rajan Jain	229- Nanoengineering Polyorganosilica (POSi) Porous Structure via Few-cycle Atomic Layer Deposition for Superior H2/CO2 Separation Author: Vinh Bui	316- Pretreatment to Enhance the Performace of Virus Filtration Author: Ranil Wickramasinghe	145- Utilizing a Green Solvent for Energy-Efficient CO2 Capture from Industrial Emission Sources via Hollow Fiber Membrane Contactors Author: Nicholas Gregorich
9:00-9:30	222- Transforming Wastewater Nitrogen into High-Value Fertilizer with Electrodialysis Author: Mohammed Tahmid	22- Combined Nanofiltration and Diafiltration for Isolation of Rare-earth Ions Author: Jun Wang	279- Fluorocarbon Reduction via Semifluorinated Copolymer Thin Films by Combining Spin Coating and Ring-Opening Metathesis Polymerization Author: Matthew Vasuta	1- Polymers of Intrinsic Microporosity: Promising materials for water desalination Author: Mahmoud Abdulhamid	267- Mechanistic Understanding and Modeling of AAV Full and Empty Capsid Separation using Multi-modal Ligands Author: Xianghong Qian	158- Machine Learning–Assisted Inverse Design and Operability Analysis of Membranes for Direct Air Capture Author: Vitor Renan Vitor Gama
9:30-10:00	430- Lack of Chemical Stability of Metal Organic Frameworks (MOFs) in Mixed Matrix Membranes for Aqueous Separation Applications Author: Matthew Davenport	119- High-throughput ion permeability analysis of 2D laminar membranes using robotic ICP platform Author: Min A Kim	338- Effect of Widespread Stacking Disorder on Single Crystal Covalent Organic Framework (COFs) Membrane Performance Author: Katie Li-Oakey	295- Leveraging dielectric exclusion to improve the desalination transport properties of hydrated polymers by engineering water/polymer interactions Author: Sean Bannon	390- Membranes for Rapid Protein Analysis Author: Merlin Bruening	198- Scaleup and Pilot Testing of a High Permeance Membrane for CO2 Capture from a Steel Mill Blast Furnace Author: David Hopkinson
10:00-10:30	29- Stable Two-Dimensional Siloxene Membranes for Molecular Separations Author: Dinesh Mahalingam	124- Engineering Cobalt Selective Membranes to Support the Energy Transition Author: Lauren Mazurowski	387- Scalable Manufacture of Loose-Nanofiltration Polymeric Membranes with Green Solvents using Slot Die Coating on a Roll-to-Roll System: Experimental and Computational Insights Author: Kwangjun Jung	188- Interfacial polymerization of micropore-generating units for molecular separation: Beyond desalination Author: Taehoon Lee	106- Utilization of Membrane Technologies for Sustainable Peptide Synthesis and Purification Author: Endras Fadhilah	325- New High Permeance, Roll to Roll Gutter Layer for Thin Film Composite Gas Separation Membranes Author: Michael McGrath
10:30-11:00	Coffee break- Starstruck Gallery					
11:00-12:00	Plenary Speech I- Symphony Ballrom					

Monday	Water Treatment II	Ion/Ion Separations II	Polymer Design/Synthesis II	NAMS Awards Session II	Modeling Gas/Vapor Transport	Carbon Capture and Sequestration II
Location	Blackbird Studio A	Blackbird Studio B	Southern Ground AB	Ocean Way	Green Room	Sound Emporium A
Session chairs	Steven Weinman	Chen Zhang	Jason Williams	Piran Kidambi	Wen Zhang	David Hopkinson
	Laura H. Arias Chavez	Stephen DeWitt	Margaret Lee	Alex Lopez	Xuemei Liang	Jason Bara
2:00-2:30	125- Harnessing Onsite Water Reuse for Sustainable Data Center Operations Author: Reema Shinh	86- Rational design of oriented mixed-matrix metal-organic framework membranes for molecular separations Author: Shuvo Datta	138- Designed for Recycling: Dynamic Crosslinking of Solvent-Resistant Membranes Author: Malinalli Ramirez Martinez	185- A channel based paradigm for membrane separations Author: Manish Kumar	150- Permeability and Selectivity Model for Separation Membranes Based on Fractal Analysis of Porous Structure Author: Riccardo Lovison	346- Effects of humidity on gas transport phenomena in PEG-based CO2 separation membranes Author: Ryan Johnson
2:30-3:00	440- Performance evaluation of commercial RO membranes in treating produced water with effective membrane cleaning approach Author: Md Hasib Al Mahbub	384- Fabrication and Performance of Bipolar Membranes with 3D Junctions Author: Peter Pintauro	312- Upcycled PVC Membranes from Waste Plastics: Evaluating the Potential for Additive Leaching (Dioctyl Phthalate) Author: ATTA UR RAZZAQ	173- Simplified Synthesis of Biomimetic Self-Assemblies for Nanofiltration and Ion-Ion Separations Author: Ronald Vogler	436- Investigating the Energetics of Condensable Gas Transport in Triptycene-Poly(arylene ether) Membranes Author: Erin Sturd	378- Beyond Power: CO2 Capture from Industrial Sources with Polaris™ Membranes Author: Jay Kniep
3:00-3:30	30- Alternating Potential Operation of Electro-Conductive Membranes: A Dual Approach to Mitigating Biofouling and Scaling in Advanced Wastewater Treatment Author: Avner Ronen	376- Overcoming the Conductance versus Crossover Trade-off in State-of-the-Art Proton Exchange Fuel-Cell Membranes by Incorporating Atomically Thin Chemical Vapor Deposition Graphene Author: Pavan Chaturvedi	167- Solvent Resistance and Adsorption Mechanisms Studies of Deep Eutectic Solvents on Polyvinylidene Fluoride Membranes Author: Odianosen Ewah	41- Bridging the Gap Between Pure and Mixed-Gas Performance of Thin-Film Composite Membranes: Advancing Cell Design and Testing for Scalable Module Development Author: Thien Tran	136- Dynamic Shifts in Gas Transport: A Framework for Discriminating Solution-Diffusion and Pore-Flow Mechanisms via Media Atom Motion Author: Jianhao Qian	424- High-Performance Water-Based Membranes for Efficient and Stable Carbon Dioxide Separation Author: Kian Lopez
3:30-4:00	165- Characterization of polymer thin films by atomic force microscopy Author: Olivia McIntee	244- Green Chemistry with Coupled Electrocatalysis and Electrodialysis: enhancing Selectivity and Conversion Efficiency Author: Kaylin Hicks	253- A Versatile Polymer Precursor for Carbon Molecular Sieve Gas Separation Membranes Author: Chen Zhang	216- Developing and Assessing the Scale-Up Life Cycle Impacts of Polysulfone Membranes with Chemically Immobilized Silver Nanoparticles and Eco-Friendly Solvents Author: David Lu	19- Decoupling the Influence of Sorption and Diffusion on Membrane Permselectivity in Pervaporation Author: Sk Md Ali Zaker Shawon	281- Atomistic Insights into the Reactive Diffusion of CO_2 in TMG-Based Facilitated Transport Membranes Author: Changlong Zou
4:00-4:30	Coffee break- Starstruck Gallery					
	Hollow Fiber and Gas Membrane Design	Dehydration/Dehumidification Separations	Hydrogen Separations	Membrane Fouling/Biofouling I	High Salinity Streams and Unconventional Systems	Biosystems and Bioreactors I
Location	Blackbird Studio A	Blackbird Studio B	Southern Ground AB	Ocean Way	Green Room	Sound Emporium A
Session chairs	Allyson McGaughey	Oishi Sanyal	Kar Chun	Swatantra Pratap Singh	Ben Sundell	Jianjia Yu
	Liren Xu	Witopo Salim	Minghui Wang	Dean Welsh	Yoram Cohen	Lauren Ward
4:30-5:00	429- Hydrogen Sulfide and Hydrocarbon Separation Properties of CTA Hollow Fiber Membranes for Natural Gas Sweetening Author: Dan Miller	191- Influence of the temperature and surface charge on the performance of nanocellulose based facilitated transport membranes for carbon capture Author: Elena Medri	59- Composite graphene oxide hollow fiber membranes for hydrogen separation Author: Miria Reis	182- Electroconductive Ultrafiltration Membranes Enhanced with Nickel-Doped Laser-Induced Graphene for Efficient Removal of Emerging Pollutant Author: Swatantra Singh	40- From Medicine to Water: Dialysis as a Novel Solution for High-Salinity Organic Wastewater Treatment Author: Yuanmiaoliang Chen	73- A Process-Structure-Property Modeling and Optimization Framework for Membrane-based Selective Nutrient Recovery Author: Md Masud
5:00-5:30	219- Identifying key precursor modification parameters for enhancing the separation performance of carbon molecular sieve (CMS) hollow fiber membranes Author: Brian Leonard	431- Engineering polymeric membranes for high-performance air dehumidification Author: Faheem Akhtar	301- Architecting High Performance H2/CO2 Separation Membranes: Synergistic Polymer Blends of m-PBI with Microporous Polymers Author: Joseph Emery	353-Co-treatment of municipal wastewater and crop harvest waste by an anaerobic membrane bioreactor for providing water, nutrients, and energy to controlled environment agriculture Author: Weiming Qi	91- High Recovery RO-NF Desalination of Groundwater of High mineral Scaling Potential Author: Yakubu Jarma	5- Water-Pore Flow Permeation through Multivalent Artificial Water Channels Author: Mihail BARBOIU
5:30-6:00	303- Polyimide Hollow Fiber Membrane Contactors for Air Humidification in Fuel Cell Systems: Optimization of Membrane Morphology and Transport Properties via Wet-Gap Spinning Author: Liang-Chih Ma	262- Synthesis and Water Vapor Separation Properties of Azine-linked COF Membranes Author: Jose Lopez-Cazares	307- Modeling Analysis of Membrane Reactor for Methane Steam Reforming for Hydrogen Production - A Comparative Study on CO2- and H2-Selective Membranes Author: Oscar Ovalle-Encinia	45- Hydrophobic electrically conducting membranes (ECMs) for enhancing ammonia recovery coupled with membrane fouling mitigation Author: SIYAO QI	230- Desalination of High Salinity Brines Using Single-Stage Osmotically Assisted Reverse Osmosis Author: Glogina Boakye	336- Fabrication of Asymmetric Selective Bilayer Membranes with Tight Pores at the Interface and Micro-pores within the Matrix using Dual-Layer Slot Die Coating on a Roll-to-Roll Author: Minwoo Jung
6:00-6:30	270- Organosoluble Pentiptycene-containing Polybenzoxazole Membranes for High Temperature Gas Separations Author: Sandra Weber	366- Innovative zeolite membrane reactors for Dimethyl ether production by MeOH dehydration Author: Giuseppe Barbieri	76- Metallic Membrane Reactors for Blue Hydrogen Production Author: Hani Hawa	89- LOCALIZED SURFACE PLASMON RESONANCE (LSPR) FOR PREDICTING FOULING AND SCALING IN RO MEMBRANES DURING SEAWATER DESALINATION Author: Iniya Prasanthi	74- Investigating the Impact of Temperature and Concentration on Membrane-Assisted Crystallization of Lithium Fluoride Solution Author: Giuseppe Prenesti	285- Biocatalytic Membrane for the Degradation of Diisononyl Phthalate Author: Hadi Rouhi
	Poster Session- Symphony Ballroom					

Tuesday	Ion Transport Investigations I	WA-MS Session	Medical Applications	Critical Mineral and Resource Recovery	Aqueous Phase Modeling	Nanofiltration and Ultrafiltration I
Location	Blackbird Studio A	Blackbird Studio B	Southern Ground AB	Ocean Way	Green Room	Sound Emporium A
Session chairs	Arif Khan	Ranil Wickramasinghe	Christina (Tina) Carbrello	Victor Kusuma	Geoff Geise	Nick Gregorich
	Meng Shen	Uwe Beuscher	Kim, Albert S.	Mahdi Malmali	Lara Heberle	Ramesh Bhave
8:00-8:30	84- Specific Ion Effects on Ion Transport in Charged Polymer Membranes Author: David Kitto	8:00-8:15 WA-MS Introduction, purpose, history Ranil Wickramasinghe	63- The application of tangential flow filtration for the continuous processing of precipitated monoclonal antibodies: process optimization in screened membrane cassettes Author: Zachary Badinger	231- Zwitterionic Membranes for Rare Earth Element (REE) Separations Author: Nafiseh Khoshnevisan	292- Interpretation of membrane transport experiments via the solution-diffusion model in the limit of low pressure Author: Kevin Reimund	69- INFLUENCE OF SURFACE PATTERNING COMBINED WITH MOF GRAFTING ON OIL FOULING AND SELENIUM REMOVAL DURING NANOFILTRATION Author: Lauren Ward
8:30-9:00	85- Theory for Dynamic Ion Transport in Ion-Shuttling-Electrodes for Electrochemical Ion Pumping Author: Weifan Liu	8:15-8:35 Guest speaker representing NAMS Membrane Technology Advances in Water and Bio are through Collaborations Dibakar (DB) Bhattacharyya	109- Biocompatible Polyacrylonitrile Hollow Fiber Membranes with Enhanced Hemodialysis Performance Author: Akshay Modi	282- Separation and Recovery of Rare Earth Elements from EV magnets using Membrane Solvent Extraction System Author: Shailesh Singh Dangwal	328- Solvent transport through swollen polymers: where do we go now? Author: Guy Ramon	201- Sustainable Polyelectrolyte Nanofiltration Membranes Author: OLAWUMI SADARE
9:00-9:30	116- Understanding the role of fixed charged group hydrophobicity on nitrate selectivity in anion-exchange membranes Author: Harsh Patel	8.35 – 8.55 Guest speaker representing EMS Large scale industrial biomimetic membranes for efficient Angstrom-scale molecular separations Mihai Barboiu	347- Patterned anion exchange membranes for abiotic, medically implanted glucose consuming fuel cells Author: Thomas DosSantos DiSorbo	443- Selective recovery of rare earth elements by membrane-based solvent extraction process Author: Md Arif Khan	226- Transport Modeling in Low-Salt-Rejection Reverse Osmosis (LSRRO) Membranes Author: Martina del Cerro	298- Nodular networks in hydrated polyamide desalination membranes enhance water transport Author: Yuexiao Shen
9:30-10:00	154- Specific Ion Selectivity Between Monovalent Cations in Commercial Cation Exchange Membranes Author: Vaishnavi Koyilath Nandakumar	8.55-9.15 Guest speaker representing AMS Collaborative Efforts for Innovative Membrane Solutions for Wastewater Treatment in Oil and Gas Operations Swatantra Singh	75- Ethanol Vapour-Induced Morphological Influences on the Water Barrier Performance of PLA Films: A Multi-component Study Author: Sean McIntyre	283- Designing Ion-Exchange Membranes for Vanadium Redox Flow Batteries Author: Gregory Reimonn	420- Methods for Measuring Experimental Energy Barriers to Transport in Salt-Rejecting Membranes Author: Mohammad Allouzi	294- Anionic covalent organic framework membranes for selective PFAS removal and dye desalination Author: Xiaoli Ma
10:00-10:30	121- Role of Frictional Forces in Salt and Water Transport through Polyamide Reverse Osmosis Membranes Author: Kevin Pataroque	9.15 – 10.30 Panel discussion of international collaborations between academia and industry Uwe Beuscher, Ayse Asatekin, Lidietta Giorno, Abaynesh Gebreyohannes, Jianxin Li	407- High-Performance Hemofiltration via Molecular Sieving and Ultra-Low Friction in Carbon Nanotube Membranes Author: Peifu Cheng	65- Membrane-Assisted Entrainer Strategy for Efficient and Sustainable Purification Processes Author: Seung Hwan Kim	379- Comparison of classical hydrodynamic models of transport through porous membranes Author: Minhao Xiao	221- XPRIZE Water Scarcity: Driving a Desalination Revolution Author: Hossein Atoufi
10:30-11:00	Coffee break- Starstruck Gallery					
11:00-12:00	Plenary Speech II- Symphony Ballrom					

Tuesday	Ion Transport Investigations II	Mixed Matrix Membranes	Modeling & Simulation of Membrane Processes	Membrane Fouling/Biofouling II	Biofilms and Biosystems Recovery	Nanofiltration and Ultrafiltration II
Location	Blackbird Studio A	Blackbird Studio B	Southern Ground AB	Ocean Way	Green Room	Sound Emporium A
Session chairs	Guy Ramon	Dr. Tequila Harris	Uwe Beuscher	Swatantra Pratap Singh	Lara Heberle	Nick Gregorich
	Arif Khan	Amirsalar R. Esfahani	Yang Han	Dean Welsh	Muskan Sonker	Ramesh Bhave
1:30-2:00	234- Beyond the First Shell: A Free Energy Outlook on Dehydration-Controlled Ion Transport in Polyamide Membranes Author: Hengyu Xu	323- Insights into the In-Situ Growth of ZIF-L Metal-Organic Framework on Ultrafiltration Membrane: MOF Growth Kinetics Effects on Water Permeability and Dye Separation Performance Author: Himangshu Mondal	300- Integrating CFD and Optimization Under Uncertainty in High-Flux Membrane Design for Carbon Capture Author: Cheick Dosso	107- Patterned Membranes for Biofouling Mitigation by Electrospinning Author: Amos Taiswa	189- Linking Biomass Properties and Membrane Fouling in an Anaerobic Membrane Bioreactor for Domestic Wastewater Treatment Author: Mor Shental-Isaacs	357- Nanofiltration using Polysulfone Membranes Author: Jian Tan
2:00-2:30	211- Ion Sorption Within Polyphenyl Alkylene Ion Exchange Membranes Author: Timothy Hudak	361- Development of highly hydrophobic self-heating Fe3O4-PVDF membranes Author: Amanda Loreti Hupsel	133- Domain Aware Modeling of Multi-Mode Operation of RO water treatment and desalination system via Graph Convolutional Networks Author: Bilal Khan	20- Fabrication of Photocatalytic TiO2-f-MWCNTs Biodegradable Polylactic acid Mixed Matrix Membrane for Enhanced Anti-Fouling Properties Author: Farah Abuhantash	95- High-Pressure Reverse Osmosis for Continuous Downstream Concentration of Bio-Based Carboxylic Acids Author: Yian Chen	426- Employing MOF-808-Functionalized NF-Membrane for Selective Extraction of Selenium from Produced Water Author: Sweta Modak
2:30-3:00	224- Mixed salt sorption in ion exchange membranes: effect of counter-ion identity Author: Aubrey Quigley	6- On the role of the porous support membrane in seawater reverse osmosis membrane synthesis, properties and performance Author: Eric Hoek	375- Engineering ion transport in polymer membranes for water purification and energy applications Author: Geoffrey Geise	322- Elucidating flowrate effects in anion-exchange membrane processes for continuous production of adenovirus therapeutics Author: Claire Velikonja	4- Synchronous electrosynthesis and separation of ammonia and chlorine from waste brinestreams via a stacked membrane-free electrolyzer Author: Wen Zhang	213- From theory to practice: Challenges in capillary flow porometry for membranes Author: Dana Dutczak
3:00-3:30	237- What is the Role of Relative Humidity on Conductivity in Polymer Electrolytes? Author: Nico Marioni	38- Fabrication and Scale-Up of Porous Polybenzimidazole (PBI) Supports for Gas Separation Composite Membranes Author: Lingxiang Zhu	9- Molecular Simulations Reveal Gas Transport Mechanisms in Polyamide Membranes Author: Ruoyu Wang	383- Sterically stable nanocrystal cellulose (SNCC) modified microfiltration membranes for the dual-enhanced NOM fouling and biofouling resistance Author: Ji Qin	370- Evaluation of hollow fiber membrane contactor reactor in reactive extraction for biodiesel production Author: Frederico Kronemberger	309- Novel Organic Solvent Nanofiltration Membranes for Hydrocarbon Separations Author: Meltem Urgun-Demirtas
3:30-4:00	Coffee break- Starstruck Gallery					
	Ion Transport Investigations III	Membrane Distillation	Electrochemical Separations I	Olefin/Paraffin Separations	Organic Solvent Separations	Biosystems and Bioreactors II
Location	Blackbird Studio A	Blackbird Studio B	Southern Ground AB	Ocean Way	Green Room	Sound Emporium A
Session chairs	Guy Ramon	William B Krantz	David Ladner	Mahmoud Abdulhamid	Haiqing Lin	Jianjia Yu
	Meng Shen	Marie-Eve Langevin	Dan Miller	Bratin Sengupta	Abhishek Mondal	Lauren Ward
4:00-4:30	296- Selective Ion Transport in Precision Polyanion Membranes Author: Kyoungmin Kim	8- Synthesis of Recyclable Microporous Polyimide-Based Membranes for Membrane Distillation Author: Toheeb Obidara	18- Enhancing Li+/Mg2+ Separation Properties by Optimizing Positive Charges of NF Membrane Surfaces Author: Sagnik Das	122- Carbon Capture and Light Paraffin Separation Membranes Based on Poly (1,3-Dioxolane)/[hmim][Tf2N] Blends Author: Justin Rosenthal	268- Organic solvent reverse osmosis membrane for cyclic siloxane purification Author: Hammed Balogun	287- The unusual transport of phenolic compounds in biorefinery streams through polymer membranes Author: Woo Jin Jang
4:30-5:00	324- Ion Transport in Microporous MFI Zeolite Membranes Author: Hao Yu	93- Enhancing the performance of vacuum membrane distillation (VMD) membrane through solvent crystallization with a bi-solvent system during phase inversion Author: Nguyen Lam	142- Bipolar membranes electrodialysis for water purification and resource recovery Author: Bin Yang	51- Advancement of Hybrid Membrane Amine-Absorption Gas Treating for Highly Sour Natural Gas Reserve Development Author: GARBA YAHAYA	249- Microporous polymer membranes for ion and organic molecules separations Author: Wong Kar Chun	266- Polymer-Silica Caged Biocatalytic Porous Media for Biofouling Control in Anaerobic Membrane Bioreactors Author: Syed Salman Ali Shah
5:00-5:30	402- Ion Selective In-plane Transport in Stacked Layers of Graphene Oxide Author: Piran Kidambi	97- Turning Aquaculture Waste into Fertilizer and Clean Water: Optimizing Membrane Distillation for N and P Recovery Author: Erica Bertozzi	210- Membrane/Filter Platforms for Anion Exchange Membranes in Difficult Aqueous Separations Author: Sam Thompson	100- Technology for Cyclic Siloxane Removal: Utilizing Hollow Fibers for Sustainable Silicone Oil Purification Author: Jinyoung Kim	13- The physical basis for solvent flow in organic solvent nanofiltration Author: Hanqing Fan	7- Texture-Driven Anti-Biofilm Efficacy of Laser-Induced Graphene: A Promising Solution to Control Membrane Biofouling Author: Kritika Jashrapuria
5:30-6:00	447- Transport of Ionic Chemotherapy Drugs in Charged Biosponge Block Copolymers Author: Hee Jeung Oh	396- Fibril contact angle: a new metric to predict wetting resistance for membrane distillation Author: Shounak Joshi	242- Affinity membranes containing complexing agents for Li+/Mg2+ separation Author: Jada Mowatt	233- Evaluation of NH3 Diffusivity and Solubility in ZnCl2 Immobilized Molten Salt (IMS) Membranes Author: Moses Adejumo	35- Design and Characterization of Inter-crosslinked Nanocomposite Membranes Exhibiting Stable Performance in Organic Solvent Separations Author: Lucas Condes	60- Clarification of natural extracts by ceramic hollow fiber membranes Author: Miria Reis

Wednesday	Biobased Separations	Composite Membranes	Biomimetic and Bioinspired Membranes	Desalination and Osmotic Processes I	Material and Module Design and Aging	Crosslinked polymers and molecular sieving membranes
Location	Blackbird Studio A	Blackbird Studio B	Southern Ground AB	Ocean Way	Melody Room	Sound Emporium A
Session chairs	Mary Laura Lind	Syed Islam	Jian James Zou	Thien Tran	Lucas Condes	Oishi Sanyal
	Ranil Wickramasinghe	Jay Liu	Chunqing Liu	Lakshmeesha Upadhyaya	Ying Labreche	Witopo Salim
8:00-8:30	314- Strategies to mitigate antifoam-induced membrane fouling. Effect of membrane chemistry and operating conditions on flux and transmission decline. Author: Federico Volpin	227- Development of Rubbery Thin Film Composite Membranes Through Interfacial Crosslinking Author: Niloofar Shirali	441- Industrial potential of nonwoven membrane chromatography adsorbers for the purification of biologics Author: Cristiana Boi	34- Rinse-Free Deposition of Molecular Layer-by-Layer (mLbL) Polyamide Reverse Osmosis Membranes Author: Samarpan Deb Majumder	36- Reducing Physical Aging of Microporous Polymer Membranes through Blending with Porous Polymer Networks Author: Lucas Condes	263- Techno-economic assessment of a carbon molecular sieve-based xylene isomer purification process Author: Hammed Balogun
8:30-9:00	432- Continuous Production of API Nanocrystals by a Porous Hollow Fiber Membrane-based Novel Anti-solvent Crystallizer Author: Kamalesh Sirkar	406- Thin Film Composite Ion Exchange Membranes by Electrospray Author: Yasmin Teixeira Bimbatti	272- Jammed Interconnected Bilayer Emulsions (JIBEs): A New Biomimetic Material for Ion-Selective Membranes Author: Aida Fica	12- Experimental Validation of the Centrifugal Reverse Osmosis (CRO) Process Author: William Krantz	332- Membrane testing at Air Liquide's Innovation Campus Delaware Author: David Hasse	134- Hyperselective carbon molecular sieve membranes for precise H2 and CO2 separation Author: Chen Zhang
9:00-9:30	417- A membrane process for the purification of extracellular vesicles from milk whey and their application as drug delivery systems Author: Cristiana Boi	27- Amphiphilic Polyelectrolyte Complex Bilayer Membranes Author: Nikan Mokarinezhad	413- Protein-Enabled Size-Selective Defect-Sealing of Atomically Thin 2D Membranes for Dialysis and Nanoscale Separations Author: Peifu Cheng	408- Direct membrane osmometry for measuring osmotic pressure of concentrated multi-electrolyte solutions Author: Jeffrey McCutcheon	56- Stereochemical Effects of Addition-Type Polynorbornenes for CO2/N2 Separation Author: Elizabeth Powell	380- Engineering thin film composite gas separation membranes using crosslinked nanoporous macrocycles Author: Joshua Moon
9:30-10:00	411- Tuning membrane properties and process conditions for preserving functional properties of biological macromolecules Author: Lidietta GIORNO	70- Membrane Preparation by Reticulate Rational Design Author: Ben Bikson	3- Sharklet® imprinted patterns on thin-film composite membranes for scaling resistance in water desalination Author: Jacalyn Morgan	196- Pore Structure, High-Throughput Membrane Development, and Compaction-Resistant Membranes for High-Pressure Reverse Osmosis Author: Jay Werber	157- Flat Sheet Membrane Modules for Gas Separation to Enhance Countercurrent Contacting Author: Glenn Lipscomb	57- Tailoring Carbon Molecular Sieve (CMS) Membranes for Challenging Gas Pair Separation Author: DEEPANJALI ROY
10:00-10:30	143- Purifying circular RNA by ultrafiltration Author: Scott Husson	421- Block Copolymer Adsorptive Membranes for PFAS Removal Author: Annabelle Li	228- Material Science Advances Towards Functionalized Membranes Author: Dibakar Bhattacharyya	455- Solid-state NMR methods for quantifying composition and crosslinking in polyamide-based thin film composite reverse osmosis membranes Author: C. M. Stafford	53- Mixed-gas performance, plasticization and aging behavior of thin film composite membranes based on hyper-crosslinked spirocyclic polymers Author: Maria Salome Tavera Villamizar	350- Carbon Molecular Sieves Prepared by Moderate-Temperature Carbonization of Phosphoric Acid-Doped Polyimide for O2/N2 Separation Author: Yang Jiao
10:30-11:00	Coffee break- Starstruck Gallery					
11:00-12:00	Plenary Speech III- Symphony Ballroom					

Wednesday	Circular Economy	Module and Systems Modeling	Electrochemical Separations II	Desalination and Osmotic Processes II	Inorganic Membranes
Location	Blackbird Studio A	Blackbird Studio B	Southern Ground AB	Ocean Way	Melody Room
Session chairs	Xiaobo Dong	Yang Han	David Ladner	Thien Tran	Caleb Funk
	Kerri Hickenbottom	Harekrushna Behera	Dan Miller	Lakshmeesha Upadhyaya	Katie Li-Oakey
1:30-2:00	335- Ultraselective ammonium recovery from wastewater: membrane processes drive materials design Author: Oded Nir	99- Deciphering the Impact of Transmembrane Pressure on Reverse Osmosis Composite Membrane Compaction and Performance Author: Jishan Wu	195- Thiol-Ene Photopolymerization for Highly Selective Anion Exchange Membranes with High Charge Concentration Author: Jocelyn Riet	252- Concentration of organic solvents by osmotically assisted reverse osmosis Author: Mengyang Hu	381- Ceramic Ion Exchange Membrane Operational Stability in Organic Solvents Author: Stephanie Candelaria
2:00-2:30	311- Green(er) Solvents for Upcycling Waste PVC into Membranes Author: ATTA UR RAZZAQ	297- Validation of high-flux plate-and-frame membrane module design for CO_2 capture: a CFD Approach Author: Cheick Dosso	225- Flow-through permeability apparatus for automated high-throughput analysis of complex solute mixtures Author: Aubrey Quigley	405- Solution-friction approximation as a robust theoretical framework for modeling low-salt-rejection reverse osmosis Author: Rayan Alghanayem	418- Scalable Bottom-Up Synthesis of Nanoporous Hexagonal Boron Nitride (h-BN) for Large-Area Atomically Thin Ceramic Membranes Author: Peifu Cheng
2:30-3:00	433- Methanol separation from poly (lactide) depolymerization products Author: Suer Kurklu Kocaoglu	255- Parametric Optimization of Layer-by-Layer Coated Ion Exchange Membranes: Insights from Computational Modeling Author: Omar Elshawarby	79- Pillared Laminar Phyllosilicate Membranes with Tunable Ion Sieving Author: Yining Liu	52- Enhancing forward osmosis efficiency: unraveling the physicochemical interactions between Layer-by-Layer coated membranes and Pluronic® as a draw solution Author: Aylin Kinik	260- Investigation of Hydrogen Permeation and Material Characterization of Pd-Ag-Y Membrane Author: Omid Jazani
3:00-3:30		269- The influence of spacer arm identity on salt transport in charged nanofiltration membranes Author: Laurianne Estrada		235- Salty no More: Explainable AI-Driven Discovery of Polyamide Desalination Membranes Author: Ying Li	

Plenary Speakers

Dr. William Koros

Professor, Roberto C. Goizueta Chair for Excellence in Chemical Engineering, and GRA Eminent Scholar in Membranes
Georgia Institute of Technology

Dr. William Koros holds the prestigious Roberto C. Goizueta Chair for Excellence in Chemical Engineering and is a GRA Eminent Scholar in Membranes. His research expertise lies in the fundamentals of sorption, diffusion, and permeation in polymers, molecular sieve carbons, and hybrid materials.

Dr. Koros's work focuses on the engineering of advanced membranes, sorbents, and barriers. His research encompasses the creation and characterization of these materials, which are pivotal for applications such as selective permeation and barrier packaging. The underlying principles of his research are based on thermodynamically controlled partitioning of penetrants, such as carbon dioxide, into membranes, sorbents, or barrier packaging layers. In polymers, this process involves the cooperative motion of the matrix to enable diffusion. In contrast, highly rigid carbon molecular sieves and zeolites rely on the relative size of pre-existing pores and the penetrant molecule, as the matrix motion is negligible.

Dr. Koros's group is at the forefront of developing materials that optimize the transport properties for specific components. For instance, they can engineer materials to act as barriers, selective membranes, or sorbents for a chosen penetrant like carbon dioxide. The group is also pioneering the formation of "mixed matrix composite" materials, blending metal-organic frameworks or other specialty components within a conventional polymer matrix. This innovative approach allows for enhanced optimization of transport properties while maintaining the processing ease associated with conventional polymers.

His research also delves into non-equilibrium thermodynamic and non-Fickian transport phenomena. His group studies the long-term conditioning effects of membranes and barriers exposed to high concentrations of certain penetrants and the protracted aging of glassy polymers, carbons, and inorganic membranes after formation or conditioning treatments. These phenomena often appear counterintuitive until one considers the expanded set of rules governing out-of-equilibrium materials.

Dr. Koros's contributions to the field of chemical engineering have established him as a leading figure in the development of advanced materials for membranes, sorbents, and barrier applications.

Dr. Beate Mueller-Tiemann

Chief Technology Officer

Cytiva

Dr. Beate Mueller-Tiemann's career has reflected the entire pharma value chain from early target discovery, lead generation, lead optimization, cell line development, chemical manufacturing and control (CMC) production and commercial manufacturing for both Biologics and Synthetic Molecule therapeutics.

She applies her extensive experience in building R&D pipelines and her unapologetic optimism in science and technology to her leadership, as her teams create and bring high-quality products and innovative manufacturing and R&D processes closer to patients. Beate joined Cytiva in early 2023, having held senior leadership positions at Sanofi and Bayer.

A former athlete who held the 800-meter record from at Fort-Le-Boeuf school district for decades, she believes in setting the pace and measuring performance. She finds inspiration in her diverse teams, when she sees the power of differing perspectives, experiences, opinions, and ideas being unleashed.

Beate earned her PhD at the Leibniz Institute of Virology in Hamburg, Germany in the field of Molecular Oncology. She holds master's degrees in biology and Biochemistry from the University of Bielefeld, Germany, and the University of Montpellier in France.

Beate is an elected member of the Board of the Dechema, German Society for Chemical Engineering and Biotechnology. She supports female talents with mentoring programs and is a member of FiDAR, an association that works towards increasing the representation of women on German corporate boards.

Dr. Winston Ho

Distinguished Professor of Engineering, Chemical and Biomolecular Engineering

The Ohio State University

Dr. W.S. Winston Ho is a Distinguished Professor of Engineering in the William G. Lowrie Department of Chemical and Biomolecular Engineering and the Department of Materials Science and Engineering at The Ohio State University.

Winston has been a member of NAMS since its formation. He served as Director of NAMS, 2005 – 2011.

Currently, he has served as Separation Engineering Section Editor, Current Opinion in Chemical Engineering and on several advisory boards including the Journal of Membrane Science. He has chaired many membrane and separations sessions at NAMS and AIChE meetings. Moreover, he gave an invited plenary lecture at the NAMS 2016 Annual Meeting, and he has developed a membrane course and taught it every year since 1999 for educating next-generation membranologists and wide-spreading membrane knowledge.

Before teaching for 25 years in 2024, he had 28 years of industrial R&D experience in membranes and separation processes. He was elected to the National Academy of Engineering, USA in 2002 in recognition of his distinguished contributions to engineering including the commercialization of gas treating technologies and the development of supported liquid membrane technology and membrane separation models. A New Jersey Inventor of the Year (1991), Winston holds more than 60 U.S. patents, generally with foreign counterparts, in membranes and separation processes.

He is a NAMS and an AIChE Fellow and has received several AIChE awards.

Oral sessions- Monday

Presenting authors are indicated by an asterisk*.

Oral 1- Water Treatment I

8:00-10:30 AM: Blackbird Studio A

Session chairs:

Steven Weinman

Laura H. Arias Chaves

8:00 AM (Blackbird Studio A, Monday)

Abstract ID- 369

Characterization, Performance Evaluation, and Toxicological assessment of Polysulfone-Sulfonated Polyether Ether Ketone Membranes for Water separation Applications

Muhammad Usman Yousaf (University of Kentucky) *, Lucca Cortarelli (University of Kentucky), Nerissa Kiprotich (University of Kentucky), Nirupam Aich (University of Nebraska – Lincoln), Olga V. Tsyusko (University of Kentucky), Isabel Escobar (University of Kentucky)

8:30 AM (Blackbird Studio A, Monday)

Abstract ID- 58

Design principles of reactive nanofiltration membranes for water treatment

Yanghua Duan (Rice University) *, Ruoyu Wang (Rice University), Amit Shocron (Yale University), Menachem Elimelech (Rice University)

9:00 AM (Blackbird Studio A, Monday)

Abstract ID- 222

Transforming Wastewater Nitrogen into High-Value Fertilizer with Electrodialysis

Mohammed Tahmid (Georgia Institute of Technology) *, Marta Hatzell (Georgia Institute of Technology)

9:30 AM (Blackbird Studio A, Monday)

Abstract ID- 430

Lack of Chemical Stability of Metal Organic Frameworks (MOFs) in Mixed Matrix Membranes for Aqueous Separation Applications

Matthew Davenport (The University of Texas at Austin) *, Kevin Reimund (The University of Texas at Austin), Matthew Dombchik (The University of Texas at Austin), Benny Freeman (The University of Texas at Austin)

10:00 AM (Blackbird Studio A, Monday)

Abstract ID- 29

Stable Two-Dimensional Siloxene Membranes for Molecular Separations

Dinesh Mahalingam (University of Bath) *

Oral 2- Ion/Ion Separations I

8:00-10:30 AM: Blackbird Studio B

Session chairs:

Hani Hawa

Joshua Moon

8:00 AM (Blackbird Studio B, Monday)

Abstract ID- 47

Poly (Terphenyl Fluorenyl Isatin) with Fluorene-Linked Sulfonate Groups and Microblock Structures for Proton Exchange Membranes

Chaoyi Ba (Argonne National Laboratory) *, Shu Xu (Argonne National Laboratory), Christopher Arges (Argonne National Laboratory), Jae Hyung Park (Argonne National Laboratory), Jinkyu Park (Argonne National Laboratory), Meltem Urgun Demirtas (Argonne National Laboratory)

8:30 AM (Blackbird Studio B, Monday)

Abstract ID- 360

Looking beyond size and charge: A roadmap to ion separations for the global energy transition

Ryan Kingsbury (Princeton University) *

9:00 AM (Blackbird Studio B, Monday)

Abstract ID- 22

Combined Nanofiltration and Diafiltration for Isolation of Rare-earth Ions

Jun Wang (University of Notre Dame) *, Dong Ding (University of Notre Dame), Laurianne Lair (University of Notre Dame), Andriy Yaroshchuk (Polytechnic University of Catalonia), William Phillip (University of Notre Dame), Merlin Bruening (University of Notre Dame)

9:30 AM (Blackbird Studio B, Monday)

Abstract ID- 119

High-throughput ion permeability analysis of 2D laminar membranes using robotic ICP platform

Min A Kim (Argonne National Laboratory) *, Ilya Shkrob (Argonne National Laboratory), Yining Liu (University of Chicago), Seth Darling (Argonne National Laboratory)

10:00 AM (Blackbird Studio B, Monday)

Abstract ID- 124

Engineering Cobalt Selective Membranes to Support the Energy Transition

Lauren Mazurowski (Yale University) *, Zhongren Jiao (Yale University), Menachem Elimelech (Rice University)

Oral 3- Polymer Design/Synthesis I

8:00-10:30 AM: Southern Ground AB

Session chairs:

Jason Williams

Margaret Lee

8:00 AM (Southern Ground AB, Monday)

Abstract ID- 64

Structure-Property-Performance Relationships of Upcycling Waste PMMA in PVDF Membrane Design

Joseph Danner (The University of Alabama) *, Liam Papier (The University of Alabama), Mia Eisenberg (The University of Alabama), Pravin Shinde (The University of Alabama), Steven Weinman (The University of Alabama)

8:30 AM (Southern Ground AB, Monday)

Abstract ID- 104

Adhesion behavior in micro/ultra-filtration membranes bonded under industry-relevant conditions

Rajan Jain (University of Colorado Boulder) *, Jaylene Martinez (University of Colorado Boulder), Riccardo Lovison (University of Colorado Boulder), Christina Carbrello (MilliporeSigma), Kathy Youngbear (MilliporeSigma), Sean Foley (MilliporeSigma), Rong Long (University of Colorado Boulder), Yifu Ding (University of Colorado Boulder)

9:00 AM (Southern Ground AB, Monday)

Abstract ID- 279

Fluorocarbon Reduction via Semifluorinated Copolymer Thin Films by Combining Spin Coating and Ring-Opening Metathesis Polymerization

Matthew Vasuta (Vanderbilt University) *, Zane Parkerson (Vanderbilt University), Tyler Oddo (Vanderbilt University), Bridget Rogers (Vanderbilt University), Kane Jennings (Vanderbilt University)

9:30 AM (Southern Ground AB, Monday)

Abstract ID- 338

Effect of Widespread Stacking Disorder on Single Crystal Covalent Organic Framework (COFs) Membrane Performance

Yun Kyung Shin (Pennsylvannia State University); Beatrice Bartolomei (Northwestern University); Caleb Johannesmeyer (University of Wyoming); Will Dichtel (Norwestern University); Adri van Duin (Pennsylvania State University); Katie Li-Oakey (University of Wyoming)*

10:00 AM (Southern Ground AB, Monday)

Abstract ID- 387

Scalable Manufacture of Loose-Nanofiltration Polymeric Membranes with Green Solvents using Slot Die Coating on a Roll-to-Roll System: Experimental and Computational Insights

Kwangjun Jung (Georgia Institute of Technology) *, David Lu (University of Kentucky), Isabel Escobar (University of

Kentucky), Tequila Harris (Georgia Institute of Technology)

Oral 4- NAMS Awards Session I

8:00-10:30 AM: Ocean Way

Session chairs:

Xiaobo Dong

Piran Kidambi

8:00 AM (Ocean Way, Monday)

Abstract ID- 344

Self-assembling, charge-containing polymers for highly selective and fouling-resistant membranes

Ayse Asatekin (Tufts)*

8:30 AM (Ocean Way, Monday)

Abstract ID- 229

Nanoengineering Polyorganosilica (POSi) Porous Structure via Few-cycle Atomic Layer Deposition for Superior H2/CO2 Separation

Vinh Bui (University at Buffalo) *, Amandine Tirino (University at Buffalo), Ameya Tandel (University at Buffalo), Erda Deng (University at Buffalo), Won Il Lee (Stony Brook University), Leiqing Hu (University at Buffalo), Lingxiang Zhu (National Energy Technology Laboratory), Chang-Yong Nam (Brookhaven National Laboratory), Haiqing Lin (University at Buffalo)

9:00 AM (Ocean Way, Monday)

Abstract ID- 1

Polymers of Intrinsic Microporosity: Promising materials for water desalination

Mahmoud Abdulhamid (King Fahd University of Petroleum and Minerals) *

9:30 AM (Ocean Way, Monday)

Abstract ID- 295

Leveraging dielectric exclusion to improve the desalination transport properties of hydrated polymers by engineering water/polymer interactions

Sean Bannon (University of Virginia) *, Rachel Fetter (University of Virginia), Geoffrey Geise (University of Virginia)

10:00 AM (Ocean Way, Monday)

Abstract ID- 188

Interfacial polymerization of micropore-generating units for molecular separation: Beyond desalination

Taehoon Lee (Sungkyunkwan University) *, Zachary P. Smith (Massachusetts Institute of Technology)

Oral 5- Virus Filtration

8:00-10:30 AM: Green Room

Session chairs:

Cristiana Boi

Gosia Chwatko

8:00 AM (Green Room, Monday)

Abstract ID- 108

Investigating the electric field driven membrane separation of biomolecules towards increased viral vector purity

Emily Ingram (University of Kentucky) *, Trupte Patel (University of Kentucky), Mara Leach (University of Kentucky), Keshav Bhanderi (University of Kentucky), Kearstin Edmonds (University of Kentucky), Rebecca Dutch (University of Kentucky), Malgorzata Chwatko (University of Kentucky)

8:30 AM (Green Room, Monday)

Abstract ID- 316

Pretreatment to Enhance the Performace of Virus Filtration

Ranil Wickramasinghe (University of Arkansas) *, Xianghong Qian (University of Arkansas), Hironobu Shirataki (Asahi Kasei Medical), Andrew Zydney (Pennsylvania State University)

9:00 AM (Green Room, Monday)

Abstract ID- 267

Mechanistic Understanding and Modeling of AAV Full and Empty Capsid Separation using Multi-modal Ligands

Xiaolei Hao (University of Arkansas), Shu-Ting Chen (University of Arkansas), Ranil Wickramasinghe (University of Arkansas), Xianghong Qian (University of Arkansas) *

9:30 AM (Green Room, Monday)

Abstract ID- 390

Membranes for Rapid Protein Analysis

Merlin Bruening (Notre Dame) *, Junyan Yang (University of Notre Dame), Weikai Cao (University of Notre Dame)

10:00 AM (Green Room, Monday)

Abstract ID- 106

Utilization of Membrane Technologies for Sustainable Peptide Synthesis and Purification

Endras Fadhilah (University of Kentucky) *, Emily Ingram (University of Kentucky), Malgorzata Chwatko (University of Kentucky)

Oral 6- Carbon Capture and Sequestration I

8:00-10:30 AM: Sound Emporium A

Session chairs:

David Hopkinson

Jason Bara

8:00 AM (Sound Emporium A, Monday)

Abstract ID- 393

Bringing MTR's Polaris™ membrane CO2 capture process to commercial scale

Richard Baker (Membrane Technology & Research Inc.), Phillip Behm (Membrane Technology & Research Inc.), Alicia Breen (Membrane Technology & Research Inc.), Brice Freeman (Membrane Technology & Research Inc.), Pingjiao Hao (Membrane Technology & Research Inc.), Jenny He (Membrane Technology & Research Inc.), Thomas Hofmann (Membrane Technology & Research Inc.), Ivy Huang (Membrane Technology & Research Inc.), Jay Kniep (Membrane Technology & Research Inc.), Tim Merkel (Membrane Technology & Research Inc.), Witopo Salim (Membrane Technology & Research Inc.) *, Ray McKaskle (Trimeric Corporation), Dana Pierik (Sargent & Lundy), Kevin Lauzze (Sargent & Lundy), William Morris (Wyoming Integrated Test Center)

8:30 AM (Sound Emporium A, Monday)

Abstract ID- 145

Utilizing a Green Solvent for Energy-Efficient CO2 Capture from Industrial Emission Sources via Hollow Fiber Membrane Contactors

Nicholas Gregorich (Oak Ridge National Laboratory) *, Mary Irwin (Oak Ridge National Laboratory), Zachary Coin (Oak Ridge National Laboratory), Stephen DeWitt (Arkema Inc), Ramesh Bhave (Oak Ridge National Laboratory), David Sholl (Oak Ridge National Laboratory), Syed Islam (Oak Ridge National Laboratory)

9:00 AM (Sound Emporium A, Monday)

Abstract ID- 158

Machine Learning–Assisted Inverse Design and Operability Analysis of Membranes for Direct Air Capture

Vitor Renan Vitor Gama (West Virginia University) *, Oishi Sanyal (West Virginia University), Fernando V. Lima (West Virginia University)

9:30 AM (Sound Emporium A, Monday)

Abstract ID- 198

Scaleup and Pilot Testing of a High Permeance Membrane for CO2 Capture from a Steel Mill Blast Furnace

David Hopkinson (NETL) *, Lingxiang Zhu (NETL), Thien Tran (NETL), Victor Kusuma (NETL), Grigorios Panagakos (NETL), Glenn Lipscomb (University of Toledo), Neil Pergar (U. S. Steel), Brenda Petrilena (U. S. Steel)

10:00 AM (Sound Emporium A, Monday)

Abstract ID- 325

New High Permeance, Roll to Roll Gutter Layer for Thin Film Composite Gas Separation Membranes

Michael McGrath (Permeance Limited and Bardo Climate Inc) *, Sam Nesbitt (Permeance Limited), Drew Letiu (Permeance Limited), Matthew Cowan (Permeance Limited), Greg Dwulet (Permeance Limited), Lingxiang Zhu (National Energy Technology Laboratory), David Hopkinson (National Energy Technology Laboratory)

Oral sessions- Monday

Presenting authors are indicated by an asterisk*.

Oral 7- Water Treatment II

2:00-4:00 PM: Blackbird Studio A

Session chairs:

Steven Weinman

Laura H. Arias Chaves

2:00 PM (Blackbird Studio A, Monday)

Abstract ID- 125

Harnessing Onsite Water Reuse for Sustainable Data Center Operations

Reema Shinh (University of Arizona) *, Leila Karimi (University of Arizona), Kerri Hickenbottom (University of Arizona)

2:30 PM (Blackbird Studio A, Monday)

Abstract ID- 440

Performance evaluation of commercial RO membranes in treating produced water with effective membrane cleaning approach

Md Hasib Al Mahbub (Texas Tech University) *, Mahdi Malmali (Texas Tech Univ

3:00 PM (Blackbird Studio A, Monday)

Abstract ID- 30

Alternating Potential Operation of Electro-Conductive Membranes: A Dual Approach to Mitigating Biofouling and Scaling in Advanced Wastewater Treatment

Avner Ronen (Ben Gurion University) *, Heyang Yuan (Temple University), Wen Zhang (NJIT), Siyao Qi (Ben Gurion University), Roy Bernstein (Ben Gurion University), Zhang Cheng (Temple University)

3:30 PM (Blackbird Studio A, Monday)

Abstract ID- 165

Characterization of polymer thin films by atomic force microscopy

Olivia McIntee (National Renewable Energy Laboratory)*; Nurshaun Sreedhar (National Renewable Energy Laboratory); Brian Welch (Israel Institute of Technology); Victor Bright (University of Colorado Boulder); Abhishek Roy (National Renewable Energy Laboratory); Mou Paul (National Renewable Energy Laboratory); Alan Greenberg (University of Colorado Boulder)

Oral 8- Ion/Ion Separations II

2:00-4:00 PM: Blackbird Studio B

Session chairs:

Chen Zhang

Stephen DeWitt

2:00 PM (Blackbird Studio B, Monday)

Abstract ID- 86

Rational design of oriented mixed-matrix metal-organic framework membranes for molecular separations

Shuvo Datta (KAUST) *

2:30 PM (Blackbird Studio B, Monday)

Abstract ID- 384

Fabrication and Performance of Bipolar Membranes with 3D Junctions

Michael McGrath (Permeance Limited and Bardo Climate Inc), Zhezhou Yang (Vanderbilt University), Ryszard Wycisk (Bardo Climate Inc), Peter Pintauro (Bardo Climate Inc) *

3:00 PM (Blackbird Studio B, Monday)

Abstract ID- 376

Overcoming the Conductance versus Crossover Trade-off in State-of-the-Art Proton Exchange Fuel-Cell Membranes by Incorporating Atomically Thin Chemical Vapor Deposition Graphene

Nicole Moehring (Bosch USA), Abdul Mansoor Basha (University of Calgary), Pavan Chaturvedi (Vanderbilt Univeristy) *, Thomas Knight (UT Knoxville), Xiaozong Fan (Vanderbilt University), Peter Pintauro (Vanderbilt University), Michael Boutilier (University of Western Ontario), Kunal Karan (University of Calgary), Piran Kidambi (Vanderbilt University)

3:30 PM (Blackbird Studio B, Monday)

Abstract ID- 244

Green Chemistry with Coupled Electrocatalysis and Electrodialysis: Enhancing Selectivity and Conversion Efficiency

Kaylin Hicks (University of Arkansas) *, Jamie Hestekin (University of Arkansas)

Oral 9- Polymer Design/Synthesis II

2:00-4:00 PM: Southern Ground AB

Session chairs:

Jason Williams

Margaret Lee

2:00 PM (Southern Ground AB, Monday)

Abstract ID- 138

Designed for Recycling: Dynamic Crosslinking of Solvent-Resistant Membranes

Malinalli Ramirez Martinez (King Abdullah University of Science and Technology) *, Sandra Aristizabal (King Abdullah University of Science and Technology), Lakshmeesha Upadhyaya (King Abdullah University of Science and Technology), Abdul-Hamid Emwas (King Abdullah University of Science and Technology), Nikos Hadjichristidis (King Abdullah University of Science and Technology), Suzana Nunes (King Abdullah University of Science and Technology)

2:30 PM (Southern Ground AB, Monday)

Abstract ID- 312

Upcycled PVC Membranes from Waste Plastics: Evaluating the Potential for Additive Leaching (Dioctyl Phthalate)

ATTA UR RAZZAQ (The University of Alabama) *, Milad Esfahani (The University of Alabama)

3:00 PM (Southern Ground AB, Monday)

Abstract ID- 167

Solvent Resistance and Adsorption Mechanisms Studies of Deep Eutectic Solvents on Polyvinylidene Fluoride Membranes

Odianosen Ewah (University of Kentucky) *, Isabel Escobar (University of Kentucky)

3:30 PM (Southern Ground AB, Monday)

Abstract ID- 253

A Versatile Polymer Precursor for Carbon Molecular Sieve Gas Separation Membranes

Ching-En Ku (University of Maryland), Gaurav Iyer (University of Maryland), Yutong Wu (University of Maryland), Chen Zhang (University of Maryland) *

Oral 10- NAMS Awards Session II

2:00-4:00 PM: Ocean Way

Session chairs:

Alex Lopez

Piran Kidambi

2:00 PM (Ocean Way, Monday)

Abstract ID- 185

A channel based paradigm for membrane separations

Manish Kumar (University of Texas at Austin) *

2:30 PM (Ocean Way, Monday)

Abstract ID- 173

Simplified Synthesis of Biomimetic Self-Assemblies for Nanofiltration and Ion-Ion Separations

Ronald Vogler (The University of Texas at Austin) *, Katherine Kimball (The University of Texas at Austin), Aída Fica (The University of Texas at Austin), Dominic Bujanos (The University of Texas at Austin), Benny Freeman (The University of Texas at Austin), Manish Kumar (The University of Texas at Austin)

3:00 PM (Ocean Way, Monday)

Abstract ID- 41

Bridging the Gap Between Pure and Mixed-Gas Performance of Thin-Film Composite Membranes: Advancing Cell Design and Testing for Scalable Module Development

Thien Tran (National Energy Technology Laboratory) *, Victor Kusuma (National Energy Technology Laboratory), David Hopkinson (National Energy Technology Laboratory), Lingxiang Zhu (National Energy Technology Laboratory)

3:30 PM (Ocean Way, Monday)

Abstract ID- 216

Developing and Assessing the Scale-Up Life Cycle Impacts of Polysulfone Membranes with Chemically Immobilized Silver Nanoparticles and Eco-Friendly Solvents

David Lu (University of Kentucky) *, Isaac Oluk (University of Kentucky), Minwoo Jung (Georgia Institute of Technology), Sophia Tseng (University of Kentucky), Diana Byrne (University of Kentucky), Tequila Harris (Georgia Institute of Technology), Isabel Escobar (University of Kentucky)

Oral 11- Modeling Gas/Vapor Transport

2:00-4:00 PM: Green Room

Session chairs:

Wen Zhang

Xuemei Liang

2:00 PM (Green Room, Monday)

Abstract ID- 150

Permeability and Selectivity Model for Separation Membranes Based on Fractal Analysis of Porous Structure

Riccardo Lovison (University of Colorado Boulder) *, Yifu Ding (University of Colorado Boulder), Christina Carbrello (MilliporeSigma)

2:30 PM (Green Room, Monday)

Abstract ID- 436

Investigating the Energetics of Condensable Gas Transport in Triptycene-Poly(arylene ether) Membranes

Erin Sturd (Massachusetts Institute of Technology) *, Zachary Smith (Massachusetts Institute of Technology)

3:00 PM (Green Room, Monday)

Abstract ID- 136

Dynamic Shifts in Gas Transport: A Framework for Discriminating Solution-Diffusion and Pore-Flow Mechanisms via Media Atom Motion

Jianhao Qian (Rice university) *, Ruoyu Wang (Rice university), Menachem Elimelech (Rice university)

3:30 PM (Green Room, Monday)

Abstract ID- 19

Decoupling the Influence of Sorption and Diffusion on Membrane Permselectivity in Pervaporation

Sk Md Ali Zaker Shawon (Vanderbilt University) *

Oral 12- Carbon Capture and Sequestration II

2:00-4:00 PM: Sound Emporium A

Session chairs:

David Hopkinson

Jason Bara

2:00 PM (Sound Emporium A, Monday)

Abstract ID- 346

Effects of humidity on gas transport phenomena in PEG-based CO2 separation membranes

Ryan Johnson (University of Florida) *, Joshua Moon (University of Florida)

2:30 PM (Sound Emporium A, Monday)

Abstract ID- 378

Beyond Power: CO2 Capture from Industrial Sources with Polaris™ Membranes

Richard Baker (Membrane Technology & Research Inc.), Brice Freeman (Membrane Technology & Research Inc.), Pingjiao Hao (Membrane Technology & Research Inc.), Jay Kniep (Membrane Technology & Research Inc.) *, Tim Merkel (Membrane Technology & Research Inc.), Robert West (Membrane Technology & Research Inc.), Kevin Lauzze (Sargent & Lundy), Dana Pierik (Sargent & Lundy)

3:00 PM (Sound Emporium A, Monday)

Abstract ID- 424

High-Performance Water-Based Membranes for Efficient and Stable Carbon Dioxide Separation

Kian Lopez (University of Colorado Boulder) *, Max Saffer-Meng (University of Colorado Boulder), Mohammad Allouzi (University of Colorado Boulder), Yukai Tomsovic (University of Colorado Boulder), Anthony Straub (University of Colorado Boulder)

3:30 PM (Sound Emporium A, Monday)

Abstract ID- 281

Atomistic Insights into the Reactive Diffusion of CO_2 in TMG-Based Facilitated Transport Membranes

ZouChanglong Zou (The Ohio State University) *, Yang Han (The Ohio State University), Li-Chiang Lin (National Taiwan University)

Oral sessions- Monday

Presenting authors are indicated by an asterisk*.

Oral 13- Hollow Fiber and Gas Membrane Design

4:30- 6:30 PM: Blackbird Studio A

Session chairs:

Allyson McGaughey

Liren Xu

4:30 PM (Blackbird Studio A, Monday)

Abstract ID- 429

Hydrogen Sulfide and Hydrocarbon Separation Properties of CTA Hollow Fiber Membranes for Natural Gas Sweetening

Dan Miller (SLB) *, Atsushi Morisato (SLB), Ed Mahley (SLB)

5:00 PM (Blackbird Studio A, Monday)

Abstract ID- 219

Identifying key precursor modification parameters for enhancing the separation performance of carbon molecular sieve (CMS) hollow fiber membranes

Brian Leonard (West Virginia University) *

5:30 PM (Blackbird Studio A, Monday)

Abstract ID- 303

Polyimide Hollow Fiber Membrane Contactors for Air Humidification in Fuel Cell Systems: Optimization of Membrane Morphology and Transport Properties via Wet-Gap Spinning

Liang-Chih Ma (Parker Hannifin) *, Stephen Telford (Parker Hannifin), Simon Padron (Parker Hannifin)

6:00 PM (Blackbird Studio A, Monday)

Abstract ID- 270

Organosoluble Pentiptycene-containing Polybenzoxazole Membranes for High Temperature Gas Separations

Sandra Weber (University of Notre Dame) *, Agboola Suleiman (University of Notre Dame), Ruilan Guo (University of Notre Dame)

Oral 14- Dehydration/Dehumidification Separations

4:30- 6:30 PM: Blackbird Studio B

Session chairs:

Oishi Sanyal

Witopo Salim

4:30 PM (Blackbird Studio B, Monday)

Abstract ID- 191

Influence of the temperature and surface charge on the performance of nanocellulose based facilitated transport membranes for carbon capture

Elena Medri (Chemical and Process Engineering Department - Alma mater studiorum Bologna) *, Marco Giacinti Baschetti (University of Bologna), Ferruccio Doghieri (University of Bologna), Stefano Rossini (ENI s.p.a.), Luca Fascia (ENI s.p.a)

5:00 PM (Blackbird Studio B, Monday)

Abstract ID- 431

Engineering polymeric membranes for high-performance air dehumidification

Faheem Akhtar (LUMS)*

5:30 PM (Blackbird Studio B, Monday)

Abstract ID- 262

Synthesis and Water Vapor Separation Properties of Azine-linked COF Membranes

Jose Lopez-Cazares (Arizona State University) *, Jerry Lin (Arizona State University), Kailong Jin (Arizona State University)

6:00 PM (Blackbird Studio B, Monday)

Abstract ID- 366

Innovative zeolite membrane reactors for Dimethyl ether production by MeOH dehydration

Giuseppe Barbieri (Consiglio Nazionale delle Ricerche - Istituto per la Tecnologia delle Membrane) *, Elisa Avruscio (Consiglio Nazionale delle Ricerche - Istituto per la Tecnologia delle Membrane), Abdu W. Sabir (Chung-ang University), Pyung Soo Lee (Chung-ang University), Adele Brunetti (Consiglio Nazionale delle Ricerche - Istituto per la Tecnologia delle Membrane)

Oral 15- Hydrogen Separations

4:30- 6:30 PM: Southern Ground AB

Session chairs:

Kar Chun

Minghui Wang

4:30 PM (Southern Ground AB, Monday)

Abstract ID- 59

Composite graphene oxide hollow fiber membranes for hydrogen separation

Guilherme Ascendino (Universidade Federal de Uberlândia), Carla Hori (Universidade Federal de Uberlândia), Miria Reis (Universidade Federal de Uberlândia) *

5:00 PM (Southern Ground AB, Monday)

Abstract ID- 301

Architecting High Performance H2/CO2 Separation Membranes: Synergistic Polymer Blends of m-PBI with Microporous Polymers

Joseph Emery (Notre Dame) *, Ruilan Guo (Notre Dame)

5:30 PM (Southern Ground AB, Monday)

Abstract ID- 307

Modeling Analysis of Membrane Reactor for Methane Steam Reforming for Hydrogen Production - A Comparative Study on CO2- and H2-Selective Membranes

Oscar Ovalle-Encinia (Universidad Autónoma Metropolitana - Iztapalapa) *, Jerry Y.S. Lin (ASU), ParthKumar Patel (ASU)

6:00 PM (Southern Ground AB, Monday)

Abstract ID- 76

Metallic Membrane Reactors for Blue Hydrogen Production

Hani Hawa (Precision Combustion, Inc.) *, Subir Roychoudhury (Precision Combustion, Inc.)

Oral 16- **Membrane Fouling/Biofouling I**

4:30- 6:30 PM: Ocean Way

Session chairs:

Swatantra Pratap Singh

Dean Welsh

4:30 PM (Ocean Way, Monday)

Abstract ID- 182-

Electroconductive Ultrafiltration Membranes Enhanced with Nickel-Doped Laser-Induced Graphene for Efficient Removal of Emerging Pollutant

Swatantra Singh (IIT BOMBAY) *; Ashish Kumar (IIT Bombay)

5:00 PM (Ocean Way, Monday)

Abstract ID- 353

Co-treatment of municipal wastewater and crop harvest waste by an anaerobic membrane bioreactor for providing water, nutrients, and energy to controlled environment agriculture

Abayomi Alayande (Clemson University); Weiming Qi (Clemson University) *; Sudeep Popat (Clemson University); David Ladner (Clemson University); Gary Amy (Clemson University)

5:30 PM (Ocean Way, Monday)

Abstract ID- 45

Hydrophobic electrically conducting membranes (ECMs) for enhancing ammonia recovery coupled with membrane fouling mitigation

SIYAO QI (Ben-Gurion University) *, Avner Ronen (Ben-Gurion University of the Negev)

6:00 PM (Ocean Way, Monday)

Abstract ID- 89

LOCALIZED SURFACE PLASMON RESONANCE (LSPR) FOR PREDICTING FOULING AND SCALING IN RO MEMBRANES DURING SEAWATER DESALINATION

Iniya Prasanthi (Ben-Gurion University of the Negev) *, Mahaveer Halakarni (Ben-Gurion University of the Negev), Moshe Herzberg (Ben Gurion U)

Oral 17- High Salinity Streams and Unconventional Systems

4:30- 6:30 PM: Green Room

Session chairs:

Ben Sundell

Yoram Cohen

4:30 PM (Green Room, Monday)

Abstract ID- 40

From Medicine to Water: Dialysis as a Novel Solution for High-Salinity Organic Wastewater Treatment

Yuanmiaoliang Chen (Rice University) *, Menachem Elimelech (Rice University), Zhangxin Wang (Guangdong University of Technology)

5:00 PM (Green Room, Monday)

Abstract ID- 91

High Recovery RO-NF Desalination of Groundwater of High mineral Scaling Potential

Yakubu Jarma (University of California, Los Angeles) *, Beau Akhavain (University of California, Los Angeles), Yoram Cohen (Professor)

5:30 PM (Green Room, Monday)

Abstract ID- 230

Desalination of High Salinity Brines Using Single-Stage Osmotically Assisted Reverse Osmosis

Glogina Boakye (Auburn University) *

6:00 PM (Green Room, Monday)

Abstract ID- 74

Investigating the Impact of Temperature and Concentration on Membrane-Assisted Crystallization of Lithium Fluoride Solution

Giuseppe Prenesti (University of Calabria) *, Mustafa Ali (University of Calabria), Alessio Caravella (University of Calabria), Francesca Macedonio (ITM-CNR), Alfredo Cassano (ITM-CNR), Elena Tocci (ITM-CNR)

Oral 18- Biosystems and Bioreactors I

4:30- 6:30 PM: Sound Emporium A

Session chairs:

Jianjia Yu

Lauren Ward

4:30 PM (Sound Emporium A, Monday)

Abstract ID- 73

A Process-Structure-Property Modeling and Optimization Framework for Membrane-based Selective Nutrient Recovery

Md Masud (West Virginia University) *, Km Piash (West Virginia University), Oishi Sanyal (West Virginia University), Yuhe Tian (West Virginia University)

5:00 PM (Sound Emporium A, Monday)

Abstract ID- 5

Water-Pore Flow Permeation through Multivalent Artificial Water Channels

Mihail BARBOIU (Institut Europeen des Membranes) *

5:30 PM (Sound Emporium A, Monday)

Abstract ID- 336

Fabrication of Asymmetric Selective Bilayer Membranes with Tight Pores at the Interface and Micro-pores within the Matrix using Dual-Layer Slot Die Coating on a Roll-to-Roll

Minwoo Jung (Georgia Institute of Technology) *; David Lu (University of Kentucky); Eesh Kulshrestha (University of Kentucky); Isabel Escobar (University of Kentucky); Tequila Harris (Georgia Institute of Technology)

6:00 PM (Sound Emporium A, Monday)

Abstract ID- 285

Biocatalytic Membrane for the Degradation of Diisononyl Phthalate

Hadi Rouhi (University of Alabama) *, River Copeland (The University of Alabama), Ryan Summers (The University of Alabama), Steven Weinman (The University of Alabama)

Oral sessions- Tuesday

Presenting authors are indicated by an asterisk*.

Oral 19- Ion Transport Investigations I

8:00-10:30 AM: Blackbird Studio A

Session chairs:

Arif Khan

Meng Shen

8:00 AM (Blackbird Studio A, Tuesday)

Abstract ID- 84

Specific Ion Effects on Ion Transport in Charged Polymer Membranes

David Kitto (University of Michigan) *, Gregory Reimonn (University of Michigan), Riley Vickers (University of Michigan), Carolina Espinoza (University of Michigan), Bryan Goldsmith (University of Michigan), Jovan Kamcev (University of Michigan)

8:30 AM (Blackbird Studio A, Tuesday)

Abstract ID- 85

Theory for Dynamic Ion Transport in Ion-Shuttling-Electrodes for Electrochemical Ion Pumping

Weifan Liu (Vanderbilt University) *, Jouke Dykstra (Wageningen University & Research), Maarten Biesheuvel (Wetsus), Longqian Xu (Vanderbilt University), Shihong Lin (Vanderbilt University)

9:00 AM (Blackbird Studio A, Tuesday)

Abstract ID- 116

Understanding the role of fixed charged group hydrophobicity on nitrate selectivity in anion-exchange membranes

Harsh Patel (University of Michigan) *, Alexander Bentley (University of Michigan), Jovan Kamcev (University of Michigan)

9:30 AM (Blackbird Studio A, Tuesday)

Abstract ID- 154

Specific Ion Selectivity Between Monovalent Cations in Commercial Cation Exchange Membranes

Vaishnavi Koyilath Nandakumar (Columbia University) *

10:00 AM (Blackbird Studio A, Tuesday)

Abstract ID- 121

Role of Frictional Forces in Salt and Water Transport through Polyamide Reverse Osmosis Membranes

Kevin Pataroque (Yale University) *, Menachem Elimelech (Rice University), Subhamoy Mahajan (University of Wisconsin Madison), Martina Del Cerro (Yale University), Jonathan Mejia (Yale University), Hanqing Fan (Rice University)

Oral 20- WA-MS Session

8:00-10:30 AM: Blackbird Studio B

Session chairs:

Ranil Wickramasinghe

Uwe Beuscher

8:00 AM (Blackbird Studio B, Tuesday)

WA-MS introduction, purpose, history; introduction to structure (currently North American Membrane Society presiding) society

Ranil Wickramasinghe, WA-MS president (University of Arkansas) *

8:15 AM (Blackbird Studio B, Tuesday)

8.15 – 8.35 Guest speaker for international collaboration (representing North American Membrane Society)

Membrane Technology Advances in Water and Bio are through Collaborations

Dibakar (DB) Bhattacharyya (University of Kentucky) *

8:35 AM (Blackbird Studio B, Tuesday)

8.35 – 8.55 Guest speaker for international collaborations (representing European Membrane Society)

Large scale industrial biomimetic membranes for efficient Angstrom-scale molecular separations

Mihai Barboiu (Institut Européen des Membranes, France) *

8:55 AM (Blackbird Studio B, Tuesday)

8.55-9.15 Guest speaker for international collaborations (representing Aseanian Membrane Society)

Collaborative Efforts for Innovative Membrane Solutions for Wastewater Treatment in Oil and Gas Operations

Swatantra Singh (Indian Institute of Technology Bombay) *

9:15 AM (Blackbird Studio B, Tuesday)

9.15 – 10.30 Panel discussion of international collaborations between academia and industry

Moderator: Uwe Beuscher, WA-MS Treasurer (WL Gore)

North American Membrane Society representative: Ayse Asatekin (Tufts University)

European Membrane Society representative: Lidietta Giorno (CNR-ITM, Istituto per la Tecnologia delle Membrane)

African Membrane Society representative: Abaynesh Gebreyohannes (Osmoses)

Aseanian Membrane Society representative: Jianxin Li (Tiangong University)

Oral 21- Medical Applications

8:00-10:30 AM: Southern Ground AB

Session chairs:

Christina (Tina) Carbrello

Kim, Albert S.

8:00 AM (Southern Ground AB, Tuesday)

Abstract ID- 63

The application of tangential flow filtration for the continuous processing of precipitated monoclonal antibodies: process optimization in screened membrane cassettes

Zach Badinger (Penn State University) *, Ali Behboudi (Penn State University), Andrew Zydney (Penn State University)

8:30 AM (Southern Ground AB, Tuesday)

Abstract ID- 109

Biocompatible Polyacrylonitrile Hollow Fiber Membranes with Enhanced Hemodialysis Performance

Pranav Krishna (Indian Institute of Science Education and Research (IISER) Bhopal India), Debasis Nayak (Indian Institute of Science Education and Research (IISER) Bhopal India), Akshay Modi (Indian Institute of Science Education and Research (IISER) Bhopal India) *

9:00 AM (Southern Ground AB, Tuesday)

Abstract ID- 347

Patterned anion exchange membranes for abiotic, medically implanted glucose consuming fuel cells

Thomas DosSantos DiSorbo (University of Colorado Denver | Anschutz) *, John Pellegrino (University of Colorado Boulder), Richard Weir (University of Colorado Denver | Anschutz)

9:30 AM (Southern Ground AB, Tuesday)

Abstract ID- 75

Ethanol Vapour-Induced Morphological Influences on the Water Barrier Performance of PLA Films: A Multi-component Study

Sean McIntyre (surface measurement systems) *, Dian Xu (Michigan State University), Samantha Adjetey (Imperial College London), Rafael Auras (Michigan State University), Daryl Williams (Surface measurement systems), Paul Iacomi (Surface measurement systems)

10:00 AM (Southern Ground AB, Tuesday)

Abstract ID- 407

High-Performance Hemofiltration via Molecular Sieving and Ultra-Low Friction in Carbon Nanotube Membranes

Peifu Cheng (Vanderbilt University) *, Nicholas Ferrell (The Ohio State University Wexner Medical Center), Carl Öberg (Lund University), Steven Buchsbaum (Lawrence Livermore National Laboratory), Melinda Jue (Lawrence Livermore National Laboratory), Sei Park (Lawrence Livermore National Laboratory), Dan Wang (The Ohio

State University Wexner Medical Center), Shuvo Roy (University of California, San Francisco), Francesco Fornasiero (Lawrence Livermore National Laboratory), William Fissell (Vanderbilt University Medical Center), Piran Kidambi (Vanderbilt University)

Oral 22- Critical Mineral and Resource Recovery

8:00-10:30 AM: Ocean Way

Session chairs:

Victor Kusuma

Mahdi Malmali

8:00 AM (Ocean Way, Tuesday)

Abstract ID- 231

Zwitterionic Membranes for Rare Earth Element (REE) Separations

Nafiseh Khoshnevisan (Tufts University) *, Ayse Asatekin (Tufts University), Ashleigh Herrera (Tufts University), Christine Duval (Case Western Reserve University), Venkat Ganesan (University of Texas at Austin)

8:30 AM (Ocean Way, Tuesday)

Abstract ID- 282

Separation and Recovery of Rare Earth Elements from EV magnets using Membrane Solvent Extraction System

Shailesh Singh Dangwal (Oak Ridge National Lab) *, Zachary Coin (Oak Ridge National Lab), Syed Islam (Oak Ridge National Lab), Ramesh Bhave (Oak Ridge National Lab)

9:00 AM (Ocean Way, Tuesday)

Abstract ID- 443

Selective recovery of rare earth elements by membrane-based solvent extraction process

Victoria Bessah (North Carolina A&T State University), Shailesh Dangwal (Oak Ridge National Laboratory), Ajaratu Cole (North Carolina A&T State University), Syed Islam (Oak Ridge National Laboratory), Shamsuuddin Ilias (North Carolina A&T State University), Ramesh Bhave (Oak Ridge National Laboratory), Md Arif Khan (North Carolina A&T State University) *

9:30 AM (Ocean Way, Tuesday)

Abstract ID- 283

Designing Ion-Exchange Membranes for Vanadium Redox Flow Batteries

Author: Gregory Reimonn *

10:00 AM (Ocean Way, Tuesday)

Abstract ID- 65

Membrane-Assisted Entrainer Strategy for Efficient and Sustainable Purification Processes

Seung Hwan Kim *

Oral 23- Aqueous Phase Modeling

8:00-10:30 AM: Green Room

Session chairs:

Geoff Geise

Lara Heberle

8:00 AM (Green Room, Tuesday)

Abstract ID- 292

Interpretation of membrane transport experiments via the solution-diffusion model in the limit of low pressure

Kevin Reimund (University of Texas at Austin) *, Benny Freeman (University of Texas at Austin)

8:30 AM (Green Room, Tuesday)

Abstract ID- 328

Solvent transport through swollen polymers: where do we go now?

Guy Ramon (Technion - Israel Institute of Technology) *

9:00 AM (Green Room, Tuesday)

Abstract ID- 226

Transport Modeling in Low-Salt-Rejection Reverse Osmosis (LSRRO) Membranes

Martina del Cerro (Yale University) *, Amit Shocron (Yale University), Hanqing Fan (Rice University), Jack Gilron (Ben-Gurion University of the Negev), Menachem Elimelech (Rice University)

9:30 AM (Green Room, Tuesday)

Abstract ID- 420

Methods for Measuring Experimental Energy Barriers to Transport in Salt-Rejecting Membranes

Mohammad Allouzi (University of Colorado Boulder) *

10:00 AM (Green Room, Tuesday)

Abstract ID- 379

Comparison of classical hydrodynamic models of transport through porous membranes

Minhao Xiao (UCLA)*

Oral 24- Nanofiltration and Ultrafiltration I

8:00-10:30 AM: Sound Emporium A

Session chairs:

Nick Gregorich

Ramesh Bhave

8:00 AM (Sound Emporium A, Tuesday)

Abstract ID- 69

Influence of Surface Patterning Combined with MOF Grafting on Oil Fouling and Selenium Removal during Nanofiltration

Lauren Ward (The University of Alabama) *, Sweta Modak (The University of Alabama), Catherine Martin (The University of Alabama), Milad Esfahani (The University of Alabama), Steven Weinman (The University of Alabama)

8:30 AM (Sound Emporium A, Tuesday)

Abstract ID- 201

Sustainable Polyelectrolyte Nanofiltration Membranes

OLAWUMI SADARE (University of Bath) *, Garyfalia Zoumpouli, (University of Bath), John Chew (University of Bath), Davide Mattia (University of Bath)

9:00 AM (Sound Emporium A, Tuesday)

Abstract ID- 298

Nodular networks in hydrated polyamide desalination membranes enhance water transport

Yuexiao Shen (Texas Tech University) *, Danyang Li (Tsinghua University), Menachem Elimelech (Rice University), Xia Huang (Tsinghua University)

9:30 AM (Sound Emporium A, Tuesday)

Abstract ID- 294

Anionic covalent organic framework membranes for selective PFAS removal and dye desalination

Thanh-Tin Nguyen (University of Wisconsin-Milwaukee), Rahul Sampat Khandge (University of Wisconsin-Milwaukee), Xiaopeng Min (University of Wisconsin-Milwaukee), Peng Hu (University of Wisconsin-Milwaukee), Ho-Kuan Yu (University of Wisconsin-Milwaukee), Hanlun Gao (University of Wisconsin-Milwaukee), Yin Wang (University of Wisconsin-Milwaukee), Xiaoli Ma (University of Wisconsin-Milwaukee) *

10:00 AM (Sound Emporium A, Tuesday)

Abstract ID- 221

XPRIZE Water Scarcity: Driving a Desalination Revolution

Hossein Atoufi (XPRIZE Foundation) *

Oral sessions- Tuesday

Presenting authors are indicated by an asterisk*.

Oral 25- Ion Transport Investigations II

2:00-4:00 PM: Blackbird Studio A

Session chairs:

Guy Ramon

Arif Khan

2:00 PM (Blackbird Studio A, Tuesday)

Abstract ID- 234

Beyond the First Shell: A Free Energy Outlook on Dehydration-Controlled Ion Transport in Polyamide Membranes

Hengyu Xu (University of Wisconsin-Madison) *, Subhamoy Mahajan (University of Wisconsin-Madison), Ying Li (University of Wisconsin-Madison)

2:30 PM (Blackbird Studio A, Tuesday)

Abstract ID- 211

Ion Sorption Within Polyphenyl Alkylene Ion Exchange Membranes

Timothy Hudak (The Pennsylvania State University) *, Christopher Gorski (The Pennsylvania State University), Enrique Gomez (The Pennsylvania State Universtiy)

3:00 PM (Blackbird Studio A, Tuesday)

Abstract ID- 224

Mixed salt sorption in ion exchange membranes: effect of counter-ion identity

Aubrey Quigley (The University of Texas at Austin) *, Jordyn Yoh (The University of Texas at Austin), Nathaniel Lynd (The University of Texas at Austin), Benny Freeman (The University of Texas)

3:30 PM (Blackbird Studio A, Tuesday)

Abstract ID- 237

What is the Role of Relative Humidity on Conductivity in Polymer Electrolytes?

Nico Marioni (The University of Texas at Austin) *, Akhila Rajesh (The University of Texas at Austin), Rahul Sujanani (University of California, Santa Barbara), Zidan Zhang (The University of Texas at Austin), Leo Gordon (University of California, Santa Barbara), Raphaële Clément (University of California, Santa Barbara), Rachel Segalman (University of California, Santa Barbara), Benny Freeman (The University of Texas at Austin), Venkat Ganesan (The University of Texas at Austin)

Oral 26- Mixed Matrix Membranes

2:00-4:00 PM: Blackbird Studio B

Session chairs:

Dr. Tequila Harris

Amirsalar R. Esfahani

2:00 PM (Blackbird Studio B, Tuesday)

Abstract ID- 323

Insights into the In-Situ Growth of ZIF-L Metal-Organic Framework on Ultrafiltration Membrane: MOF Growth Kinetics Effects on Water Permeability and Dye Separation Performance

Himangshu Mondal (The University of Alabama) *, Milad Esfahani (The University of Alabama)

2:30 PM (Blackbird Studio B, Tuesday)

Abstract ID- 361

Development of highly hydrophobic self-heating Fe3O4-PVDF membranes

Amanda Loreti Hupsel (COPPE/PEQ/UFRJ) *, Cristiano Piacsek Borges (COPPE/PEQ/UFRJ), Fabiana Valéria da Fonseca (EQ/UFRJ)

3:00 PM (Blackbird Studio B, Tuesday)

Abstract ID- 6

On the role of the porous support membrane in seawater reverse osmosis membrane synthesis, properties and performance

Eric Hoek (UCLA)*

3:30 PM (Blackbird Studio B, Tuesday)

Abstract ID- 38

Fabrication and Scale-Up of Porous Polybenzimidazole (PBI) Supports for Gas Separation Composite Membranes

Lingxiang Zhu (National Energy Technology Laboratory) *, Fangming Xiang (National Energy Technology Laboratory), Thien Tran (National Energy Technology Laboratory), Victor Kusuma (National Energy Technology Laboratory), David Hopkinson (National Energy Technology Laboratory)

Oral 27- Modeling & Simulation of Membrane Processes

2:00-4:00 PM: Southern Ground AB

Session chairs:

Uwe Beuscher

Yang Han

2:00 PM (Southern Ground AB, Tuesday)

Abstract ID- 300

Integrating CFD and Optimization Under Uncertainty in High-Flux Membrane Design for Carbon Capture

Cheick Dosso (Carnegie Mellon University) *, Hector Pedrozo (Carnegie Mellon University), Lorenz Biegler (Carnegie Mellon University), Grigorios Panagakos (Carnegie Mellon University)

2:30 PM (Southern Ground AB, Tuesday)

Abstract ID- 133

Domain Aware Modeling of Multi-Mode Operation of RO water treatment and desalination system via Graph Convolutional Networks

Michael Clement (California State University, San Bernardino), Nathan Kelly (California State University, San Bernardino), Nora Marki (University of California, Los Angeles), Maria Soto (University of California, Los Angeles), Yoram Cohen (Professor), Bilal Khan (California State University, San BErnardino) *

3:00 PM (Southern Ground AB, Tuesday)

Abstract ID- 375

Engineering ion transport in polymer membranes for water purification and energy applications

Geoffrey Geise (University of Virginia) *

3:30 PM (Southern Ground AB, Tuesday)

Abstract ID- 9

Molecular Simulations Reveal Gas Transport Mechanisms in Polyamide Membranes

Ruoyu Wang (Rice University) *, Jianhao Qian (Rice University), Menachem Elimelech (Rice University)

Oral 28- Membrane Fouling/Biofouling II

2:00-4:00 PM: Ocean Way

Session chairs:

Swatantra Pratap Singh

Dean Welsh

2:00 PM (Ocean Way, Tuesday)

Abstract ID- 107

Patterned Membranes for Biofouling Mitigation by Electrospinning

Amos Taiswa (Montana Tech) *

2:30 PM (Ocean Way, Tuesday)

Abstract ID- 20

Fabrication of Photocatalytic TiO2-f-MWCNTs Biodegradable Polylactic acid Mixed Matrix Membrane for Enhanced Anti-Fouling Properties

Farah Abuhantash (Khalifa University) *

3:00 PM (Ocean Way, Tuesday)

Abstract ID- 322

Elucidating flowrate effects in anion-exchange membrane processes for continuous production of adenovirus therapeutics

Claire Velikonja (McMaster University) *; William Pihlainen-Bleecker (McMaster University); Cindy Shu (McMaster University); Adrian Foell (McMaster University); Landon Steenbakkers (McMaster University); Brandon Corbett (Sartorius); Chris McCready (Sartorius); Prashant Mhaskar (McMaster University); David Latulippe (McMaster University)

3:30 PM (Ocean Way, Tuesday)

Abstract ID- 383

Sterically stable nanocrystal cellulose (SNCC) modified microfiltration membranes for the dual-enhanced NOM fouling and biofouling resistance

Ji Qin (University of Minnesota) *, Boya Xiong (University of Minnesota)

Oral 29- Biofilms and Biosystems Recovery

2:00-4:00 PM: Green Room

Session chairs:

Lara Heberle

Muskan Sonker

2:00 PM (Green Room, Tuesday)

Abstract ID- 189

Linking Biomass Properties and Membrane Fouling in an Anaerobic Membrane Bioreactor for Domestic Wastewater Treatment

Mor Shental-Isaacs (Ben Gurion University of the Negev) *, Moshe Herzberg (Ben-Gurion University of the Negev), Ronit Bitton (Ben-Gurion University of the Negev), Michal Zaiden (Ben-Gurion University of the Negev)

2:30 PM (Green Room, Tuesday)

Abstract ID- 95

High-Pressure Reverse Osmosis for Continuous Downstream Concentration of Bio-Based Carboxylic Acids

Yian Chen (KGI) *, Hakan Olcay (NREL), Eric Tan (NREL), Gregg Beckham (NREL)

3:00 PM (Green Room, Tuesday)

Abstract ID- 4

Synchronous electrosynthesis and separation of ammonia and chlorine from waste brinestreams via a stacked membrane-free electrolyzer

Wen Zhang (New Jersey Institute of Technology) *

3:30 PM (Green Room, Tuesday)

Abstract ID- 370

Evaluation of hollow fiber membrane contactor reactor in reactive extraction for biodiesel production

Frederico Kronemberger (PEQ/COPPE/UFRJ) *, Nevardo Yaya (PEQ/COPPE/UFRJ), Claudio Habert (PEQ/COPPE/UFRJ)

Oral 30- Nanofiltration and Ultrafiltration II

2:00-4:00 PM: Sound Emporium A

Session chairs:

Nick Gregorich

Ramesh Bhave

2:00 PM (Sound Emporium A, Tuesday)

Abstract ID- 357

Nanofiltration using Polysulfone Membranes

Jian Tan (Pall Corporation) *, Varun Vakharia (Pall Corporation)

2:30 PM (Sound Emporium A, Tuesday)

Abstract ID- 426

Employing MOF-808-Functionalized NF-Membrane for Selective Extraction of Selenium from Produced Water

Sweta Modak (The University of Alabama) *, Milad Esfahani (The University of Alabama)

3:00 PM (Sound Emporium A, Tuesday)

Abstract ID- 213

From theory to practice: Challenges in capillary flow porometry for membranes

Dana Dutczak (POROMETER) *, Weronika Motyka (IB-FT GmbH), Loubaba Sharaf Aldeen (IB-FT GmbH), Karolina Chojnacka (IB-FT GmbH), Izabela Struzynska-Piron (POROMETER), Marina Angesleva (POROMETER), Eline Pattyn (POROMETER), Anna Sobolewska (IB-FT GmbH)

3:30 PM (Sound Emporium A, Tuesday)

Abstract ID- 309

Novel Organic Solvent Nanofiltration Membranes for Hydrocarbon Separations

Meltem Urgun-Demirtas (Argonne National Laboratory) *, Lauren valentino (Argonne National Laboratory), Chaoyi Ba (Argonne National Laboratory)

Oral sessions- Tuesday

Presenting authors are indicated by an asterisk*.

Oral 31- Ion Transport Investigations III

4:00- 6:00 PM: Blackbird Studio A

Session chairs:

Guy Ramon

Meng Shen

4:00 PM (Blackbird Studio A, Tuesday)

Abstract ID- 296

Selective Ion Transport in Precision Polyanion Membranes

Kyoungmin Kim (Florida State University) *, Chinnadurai Satheeshkumar (Florida State University), Sebastian Castro (Florida State University), Joseph Schlenoff (Florida State University), Justin Kennemur (Florida State University), Daniel Hallinan (Florida State University)

4:30 PM (Blackbird Studio A, Tuesday)

Abstract ID- 324

Ion Transport in Microporous MFI Zeolite Membranes

Hao Yu (Arizona State University) *, Lance Lai (Arizona State University), Jerry Lin (Arizona State University)

5:00 PM (Blackbird Studio A, Tuesday)

Abstract ID- 402

Ion Selective In-plane Transport in Stacked Layers of Graphene Oxide

JUMI DEKA (VANDERBILT UNIVERSITY), PAVAN CHATURVEDI (VANDERBILT UNIVERSITY), PIRAN KIDAMBI (VANDERBILT UNIVERSITY) *

5:30 PM (Blackbird Studio A, Tuesday)

Abstract ID- 447

Transport of Ionic **Chemotherapy Drugs in Charged Biosponge Block Copolymers**

Hee Jeung Oh (Pennsylvania State University) *

Oral 32- Membrane Distillation

4:00- 6:00 PM: Blackbird Studio B

Session chairs:

William B Krantz

Marie-Eve Langevin

4:00 PM (Blackbird Studio B, Tuesday)

Abstract ID- 8

Synthesis of Recyclable Microporous Polyimide-Based Membranes for Membrane Distillation

Toheeb Obidara (Northeastern University) *, Turki Baroud (King Fahd University of Petroleum Resources)

4:30 PM (Blackbird Studio B, Tuesday)

Abstract ID- 93

Enhancing the performance of vacuum membrane distillation (VMD) membrane through solvent crystallization with a bi-solvent system during phase inversion

Nguyen Lam (Nanyang Technological University) *, Paulo Jorge Da Silva Bartolo (Nanyang Technological University), Chong Tzyy Haur (Nanyang Technological University)

5:00 PM (Blackbird Studio B, Tuesday)

Abstract ID- 97

Turning Aquaculture Waste into Fertilizer and Clean Water: Optimizing Membrane Distillation for N and P Recovery

Erica Bertozzi (Politecnico di Torino) *, Gabriele Copetti (Politecnico di Torino), Lorenzo Craveri (Politecnico di Torino), Ali Naeimi Tabasian (Politecnico di Torino), Marco Malaguti (Politecnico di Torino), Alberto Tiraferri (Politecnico di Torino)

5:30 PM (Blackbird Studio B, Tuesday)

Abstract ID- 396

Fibril contact angle: a new metric to predict wetting resistance for membrane distillation

Shounak Joshi (University of Southern California) *, Allyson McGaughey (University of New Mexico), Amy Childress (University of Southern California)

Oral 33- Electrochemical Separations I

4:00- 6:00 PM: Southern Ground AB

Session chairs:

David Ladner

Dan Miller

4:00 PM (Southern Ground AB, Tuesday)

Abstract ID- 18

Enhancing Li+/Mg2+ Separation Properties by Optimizing Positive Charges of NF Membrane Surfaces

Sagnik Das (University at Buffalo) *, Haiqing Lin (University at Buffalo), Erda Deng (University at Buffalo), Kai Chen (University at buffalo), Jada Mowatt (University at Buffalo)

4:30 PM (Southern Ground AB, Tuesday)

Abstract ID- 142

Bipolar membranes electrodialysis for water purification and resource recovery

Bin Yang (National Renewable Energy Laboratory) *, Caleb Reese (National Renewable Energy Laboratory), Nurshaun Sreedhar (National Renewable Energy Laboratory; Department of Civil, Architectural and Environmental Engineering, The University of Texas at Austin), Chance Parrish (National Renewable Energy Laboratory), Mou Paul (National Renewable Energy Laboratory), Manish Kumar (Department of Civil, Architectural and Environmental Engineering, The University of Texas at Austin), Abhishek Roy (National Renewable Energy Laboratory)

5:00 PM (Southern Ground AB, Tuesday)

Abstract ID- 210

Membrane/Filter Platforms for Anion Exchange Membranes in Difficult Aqueous Separations

Sam Thompson (University of Kentucky) *, Dibakar Bhattacharyya (University of Kentucky), Jennifer Bukowski (University of Kentucky), Mattie Brock (University of Kentucky)

5:30 PM (Southern Ground AB, Tuesday)

Abstract ID- 242

Affinity membranes containing complexing agents for Li+/Mg2+ separation

Jada Mowatt (University at Buffalo, The State University of New York) *, Sagnik Das (University at Buffalo, The State University of New York), Haiqing Lin (University at Buffalo, The State University of New York)

Oral 34- Olefin/Paraffin Separations

4:00- 6:00 PM: Ocean Way

Session chairs:

Mahmoud Abdulhamid

Bratin Sengupta

4:00 PM (Ocean Way, Tuesday)

Abstract ID- 122

Carbon Capture and Light Paraffin Separation Membranes Based on Poly(1,3-Dioxolane)/[hmim][Tf2N] Blends

Justin Rosenthal (The University of Texas at Austin) *, Xavier Gatica (The University of Texas at Austin), Antoine Chamoun-Farah (The University of Texas at Austin), Adam Mann (The University of Texas at Austin), Tangqiumei Song (The University of Texas at Austin), Isaac Tan (The University of Texas at Austin), Enrique Loredo (The University of Texas at Austin), Nathaniel Lynd (The University of Texas at Austin), Gabriel Sanoja (The University of Texas at Austin), Joan Brennecke (The University of Texas at Austin), Benny Freeman (The University of Texas at Austin)

4:30 PM (Ocean Way, Tuesday)

Abstract ID- 51

Advancement of Hybrid Membrane Amine-Absorption Gas Treating for Highly Sour Natural Gas Reserve Development

GARBA YAHAYA (SAUDI ARAMCO) *, FERAS HAMAD (SAUDI ARAMCO), SEBASTIEN DUVAL (SAUDI ARAMCO)

5:00 PM (Ocean Way, Tuesday)

Abstract ID- 100

Technology for Cyclic Siloxane Removal: Utilizing Hollow Fibers for Sustainable Silicone Oil Purification

Jinyoung Kim (Georgia Tech) *, Jay Rose (Dow Chemical Company), Dimi Katsoulis (Dow Chemical Company), Reza Panah (Dow Chemical Company), Ryan Lively (Georgia Tech), William Koros (Georgia Tech)

5:30 PM (Ocean Way, Tuesday)

Abstract ID- 233

Evaluation of NH3 Diffusivity and Solubility in ZnCl2 Immobilized Molten Salt (IMS) Membranes

Moses Adejumo (Clarkson University) *, Abdulkadir Zabi (Clarkson University), Simona Liguori (Clarkson University)

Oral 35- Organic Solvent Separations

4:00- 6:00 PM: Green Room

Session chairs:

Haiqing Lin

Abhishek Mondal

4:00 PM (Green Room, Tuesday)

Abstract ID- 268

Organic solvent reverse osmosis membrane for cyclic siloxane purification

Hammed Balogun (Georgia Institute of Technology) *, Reza Haghpanah (The Dow Chemical Company), Dimitris Katsouli (The Dow Chemical Company), Jay Rose (The Dow Chemical Company), William Koros (Georgia Institute of Technology), Ryan Lively (Georgia Institute of Technology)

4:30 PM (Green Room, Tuesday)

Abstract ID- 249

Microporous polymer membranes for ion and organic molecules separations

Wong Kar Chun (National University of Singapore) *

5:00 PM (Green Room, Tuesday)

Abstract ID- 13

The physical basis for solvent flow in organic solvent nanofiltration

Hanqing Fan (Rice University) *, Jinlong He (University of Wisconsin-Madison), Dariush M. Heiranian (North Carolina State University), Weiyi Pan (Rice University), Ying Li (University of Wisconsin-Madison), Menachem Elimelech (Rice University)

5:30 PM (Green Room, Tuesday)

Abstract ID- 35

Design and Characterization of Inter-crosslinked Nanocomposite Membranes Exhibiting Stable Performance in Organic Solvent Separations

Lucas Condes (University of Oklahoma) *, Michele Galizia (University of Oklahoma)

Oral 36- Biosystems and Bioreactors II

4:00- 6:00 PM: Sound Emporium A

Session chairs:

Jianjia Yu

Lauren Ward

4:00 PM (Sound Emporium A, Tuesday)

Abstract ID- 287

The unusual transport of phenolic compounds in biorefinery streams through polymer membranes

Woo Jin Jang (Georgia Institute of Technology) *, Ryan Lively (Georgia Institute of Technology)

4:30 PM (Sound Emporium A, Tuesday)

Abstract ID- 266

Polymer-Silica Caged Biocatalytic Porous Media for Biofouling Control in Anaerobic Membrane Bioreactors

Syed Salman Ali Shah (Beijing Normal University at Zhuhai) *

5:00 PM (Sound Emporium A, Tuesday)

Abstract ID- 7

Texture-Driven Anti-Biofilm Efficacy of Laser-Induced Graphene: A Promising Solution to Control Membrane Biofouling

Kritika Jashrapuria (Indian Institute of Technology Bombay) *, Swatantra Pratap Singh (Indian Institute of Technology bombay)

5:30 PM (Sound Emporium A, Tuesday)

Abstract ID- 60

Clarification of natural extracts by ceramic hollow fiber membranes

Kristopher Dorneles (Universidade Federal de Uberlândia), Karolyna Miguel (Universidade Federal de Uberlândia), Vicelma Cardoso (Universidade Federal de Uberlândia), Miria Reis (Universidade Federal de Uberlândia) *

Oral sessions- Wednesday

Presenting authors are indicated by an asterisk*.

Oral 37- Biobased Separations

8:00-10:30 AM: Blackbird Studio A

Session chairs:

Mary Laura Lind

Ranil Wickramasinghe

8:00 AM (Blackbird Studio A, Wednesday)

Abstract ID- 314

Strategies to mitigate antifoam-induced membrane fouling. Effect of membrane chemistry and operating conditions on flux and transmission decline

Federico Volpin (Novo Nordisk) *, Bianca Schroth (Novo Nordisk)

8:30 AM (Blackbird Studio A, Wednesday)

Abstract ID- 432

Continuous Production of API Nanocrystals by a Porous Hollow Fiber Membrane-based Novel Anti-solvent Crystallizer

Yufeng Song (New Jersey Institute of Technology), Lixin Feng (New Jersey Institute of Technology), Wei Zhong (New Jersey Institute of Technology), Siddharth Tripathi (New Jersey Institute of Technology), Padmini Ananthoji (Amgen), Piero Armenante (New Jersey Institute of Technology), Kamalesh Sirkar (New Jersey Institute of Technology) *

9:00 AM (Blackbird Studio A, Wednesday)

Abstract ID- 417

A membrane process for the purification of extracellular vesicles from milk whey and their application as drug delivery systems

Sara Giancaterino (Università di Bologna), Caterina Piunti (Università di Padova), Riccardo Onesti (Università di Bologna), Elisa Cimetta (Università di Padova), Cristiana Boi (Università di Bologna) *

9:30 AM (Blackbird Studio A, Wednesday)

Abstract ID- 411

Tuning membrane properties and process conditions for preserving functional properties of biological macromolecules

Lidietta GIORNO (National Research Council of Italy - Institute on Membrane Technology (CNR-ITM)) *

10:00 AM (Blackbird Studio A, Wednesday)

Abstract ID- 143

Purifying circular RNA by ultrafiltration

Karen Guillen Cuevas (Clemson University), Xiaoming Lu (Clemson University), Marc Birtwistle (Clemson University), Scott Husson (Clemson University) *

Oral 38- Composite Membranes

8:00-10:30 AM: Blackbird Studio B

Session chairs:

Syed Islam

Jay Liu

8:00 AM (Blackbird Studio B, Wednesday)

Abstract ID- 227

Development of Rubbery Thin Film Composite Membranes Through Interfacial Crosslinking

Niloofar Shirali (University of Toronto) *, Dean Stipanic (University of Toronto), Jay Werber (university of toronto)

8:30 AM (Blackbird Studio B, Wednesday)

Abstract ID- 406

Thin Film Composite Ion Exchange Membranes by Electrospray

Yasmin Teixeira Bimbatti (UCONN) *, Deniz Ipekçi (UCONN), Xingyu Wang (UCONN), Edward Wazer (UCONN), Inara Oliveira (University of Virginia), Anna Harris (University of Virginia), Hazel Gerber (Georgia Institute of Technology), Baikun Li (UCONN), Geoffrey Geise (University of Virginia), Paul Kohl (Georgia Institute of Technology), Jeffrey R. McCutcheon (UCONN)

9:00 AM (Blackbird Studio B, Wednesday)

Abstract ID- 27

Amphiphilic Polyelectrolyte Complex Bilayer Membranes

Nikan Mokarinezhad (Tufts University) *, Luca Mazzaferro (Tufts University), Usama Sheikh (University of Connecticut), Benjamin J. Sundell (ZwitterCo, Inc.), Jeffrey R. McCutcheon (University of Connecticut), Ayse Asatekin (Tufts University)

9:30 AM (Blackbird Studio B, Wednesday)

Abstract ID- 70

Membrane Preparation by Reticulate Rational Design

Ben Bikson (Avanpore) *, Ninad Khadse (Avanpore, LLC), Sagiv Weintraub (Chromodynamix, LLC)

10:00 AM (Blackbird Studio B, Wednesday)

Abstract ID- 421

Block Copolymer Adsorptive Membranes for PFAS Removal

Annabelle Li (University of Notre Dame) *, Siwei Gu (University of Notre Dame), Kyle Doudrick (University of Notre Dame), William Phillip (University of Notre Dame)

Oral 39- Biomimetic and Bioinspired Membranes

8:00-10:30 AM: Southern Ground AB

Session chairs:

Jian James Zou

Chunqing Liu

8:00 AM (Southern Ground AB, Wednesday)

Abstract ID- 441

Industrial potential of nonwoven membrane chromatography adsorbers for the purification of biologics

Cristiana Boi (Università di Bologna) *, Jinxin Fan (North Carolina State University), Joseph Lavoie (North Carolina State University), Luca Masciolini (North Carolina State University), Riccardo Onesti (Università di Bologna)

8:30 AM (Southern Ground AB, Wednesday)

Abstract ID- 272

Jammed Interconnected Bilayer Emulsions (JIBEs): A New Biomimetic Material for Ion-Selective Membranes

Aida Fica (University of Texas) *

9:00 AM (Southern Ground AB, Wednesday)

Abstract ID- 413

Protein-Enabled Size-Selective Defect-Sealing of Atomically Thin 2D Membranes for Dialysis and Nanoscale Separations

Peifu Cheng (Vanderbilt University) *, Nicholas Ferrell (The Ohio State University Wexner Medical Center), Saban Hus (Oak Ridge National Laboratory), Nicole Moehring (Vanderbilt University), Matthew Coupin (The University of Texas at Austin), Jamie Warner (The University of Texas at Austin), An-Ping Li (Oak Ridge National Laboratory), William Fissell (Vanderbilt University Medical Center), Piran Kidambi (Vanderbilt University)

9:30 AM (Southern Ground AB, Wednesday)

Abstract ID- 3

Sharklet® imprinted patterns on thin-film composite membranes for scaling resistance in water desalination

Jacalyn Morgan (University of Colorado Boulder) *, Dhairya Agnihotry (University of Colorado Boulder), Caleb Song (University of Colorado Boulder), Keiran Fung (University of Colorado Boulder), John Pellegrino (University of Colorado Boulder)

10:00 AM (Southern Ground AB, Wednesday)

Abstract ID- 228

Material Science Advances Towards Functionalized Membranes

Dibakar Bhattacharyya (U. Kentucky) *, S Thompson (University of Kentucky), R. Mills (University of Kentucky), F. Leniz

(University of Kentucky), M. Leach
(University of Kentucky), M. Brock
(University of Kentucky), J. Bukowski
(University of Kentucky), S. Bosu
(University of Kentucky), Noah Meeks
(Southern Company)

Oral 40- Desalination and Osmotic Processes I

8:00-10:30 AM: Ocean Way

Session chairs:

Thien Tran

Lakshmeesha Upadhyaya

8:00 AM (Ocean Way, Wednesday)

Abstract ID- 34

Rinse-Free Deposition of Molecular Layer-by-Layer (mLbL) Polyamide Reverse Osmosis Membranes

Samarpan Deb Majumder (George Washington University) *, Christopher Stafford (National Institute of Standards and Technology), Xitong Liu (George Washington University)

8:30 AM (Ocean Way, Wednesday)

Abstract ID- 12

Experimental Validation of the Centrifugal Reverse Osmosis (CRO) Process

William Krantz (University of Colorado at Boulder) *, Lee Nuang Sim (Nanyang Technological University), Tzyy Haur Chong (Nanyang Technological University)

9:00 AM (Ocean Way, Wednesday)

Abstract ID- 408

Direct membrane osmometry for measuring osmotic pressure of concentrated multi-electrolyte solutions

Jeffrey McCutcheon (University of Connecticut) *, Mi Zhang (University of Connecticut), Edward Wazer (University of Connecticut), Wajeha Tauqira (University of Connecticut), George Bollas (University of Connecticut)

9:30 AM (Ocean Way, Wednesday)

Abstract ID- 196

Pore Structure, High-Throughput Membrane Development, and Compaction-Resistant Membranes for High-Pressure Reverse Osmosis

Sima Zeinali Danalou (University of Toronto), Niher Sarker (University of Toronto), Hongchen Wang (University of Toronto), Patrick Lee (University of Toronto), Jason Hattrick-Simpers (University of Toronto), Jay Werber (University of Toronto) *

10:00 AM (Ocean Way, Wednesday)

Abstract ID- 455

Solid-state NMR methods for quantifying composition and crosslinking in polyamide-based thin film composite reverse osmosis membranes

C. M. Stafford (NIST) *

Oral 41- Material and Module Design and Aging

8:00-10:30 AM: Melody Room

Session chairs:

Lucas Condes

Ying Labreche

8:00 AM (Melody Room, Wednesday)

Abstract ID- 36

Reducing Physical Aging of Microporous Polymer Membranes through Blending with Porous Polymer Networks

Lucas Condes (University of Oklahoma) *, Matthew Webb (University of Oklahoma), Tran Le (University of Oklahoma), Will Box (Air Products), Cara Doherty (CSIRO), Aditi Gali (University of Oklahoma), Leoncio Garrido (ICTP-CSIC), Jing Deng (ALTR, FLTR Inc.), Laura Matesanz (ICTP-CSIC), Angel Lozano (ICTP-CSIC), Cristina Alvarez (ICTP-CSIC), Marco Nardelli (University of North Texas), Alberto Striolo (University of Oklahoma), Anita Hill (CSIRO), Michele Galizia (University of Oklahoma)

8:30 AM (Melody Room, Wednesday)

Abstract ID- 332

Membrane testing at Air Liquide's Innovation Campus Delaware

David Hasse (Air Liquide) *, Marcos Da Conceicao (Air Liquide), Sandeep Karode (Air Liquide), Ramy Swaiden (Air Liquide), Jonathan Leister (Air Liquide), Jake Stephan (Air Liquide), Greg Myrick (Air Liquide)

9:00 AM (Melody Room, Wednesday)

Abstract ID- 56

Stereochemical Effects of Addition-Type Polynorbornenes for CO2/N2 Separation

Elizabeth Powell (University of Tennessee) *, Rahil Parikh (University of Tennessee), Brian Long (University of Tennessee)

9:30 AM (Melody Room, Wednesday)

Abstract ID- 157

Flat Sheet Membrane Modules for Gas Separation to Enhance Countercurrent Contacting

Hamed Rahnema (University of Toledo), Kale Harbick (USDA), Glenn Lipscomb (University of Toledo, Toledo) *

10:00 AM (Melody Room, Wednesday)

Abstract ID- 53

Mixed-gas performance, plasticization and aging behavior of thin film composite membranes based on hyper-crosslinked spirocyclic polymers

Maria Salome Tavera Villamizar (The University of Oklahoma) *, Lucas C. Condes (The University of Oklahoma), Matthew Webb (The University of Oklahoma), Michele Galizia (University of Oklahoma)

Oral 42- Crosslinked polymers and molecular sieving membranes

8:00-10:30 AM: Sound Emporium A

Session chairs:

Oishi Sanyal

Witopo Salim

8:00 AM (Sound Emporium A, Wednesday)

Abstract ID- 263

Techno-economic assessment of a carbon molecular sieve-based xylene isomer purification process

Hammed Balogun (Georgia Institute of Technology) *, Conrad Roos (Georgia Institute of Technology), Ryan Lively (Georgia Institute of Technology)

8:30 AM (Sound Emporium A, Wednesday)

Abstract ID- 134

Hyperselective carbon molecular sieve membranes for precise H2 and CO2 separation

Gaurav Iyer (University of Maryland), Ching-En Ku (University of Maryland), Chen Zhang (University of Maryland) *

9:00 AM (Sound Emporium A, Wednesday)

Abstract ID- 380

Engineering thin film composite gas separation membranes using crosslinked nanoporous macrocycles

Joshua Moon (University of Florida) *, Woochul Kwak (University of Florida), Madeline Howell (University of Florida), Kaitlin Slicker (University of Florida), Ryan Johnson (University of Florida), Austin Evans (University of Florida)

9:30 AM (Sound Emporium A, Wednesday)

Abstract ID- 57

Tailoring Carbon Molecular Sieve (CMS) Membranes for Challenging Gas Pair Separation

DEEPANJALI ROY (WEST VIRGINIA UNIVERSITY) *, AIDAN GHEE (WEST VIRGINIA UNIVERSITY), OISHI SANYAL (WEST VIRGINIA UNIVERSITY)

10:00 AM (Sound Emporium A, Wednesday)

Abstract ID- 350

Carbon Molecular Sieves Prepared by Moderate-Temperature Carbonization of Phosphoric Acid-Doped Polyimide for O2/N2 Separation

Yang Jiao (University at Buffalo) *, Haiqing Lin (University at Buffalo)

Oral sessions- Wednesday

Presenting authors are indicated by an asterisk*.

Oral 43- Circular Economy

1:30-3:30 PM: Blackbird Studio A

Session chairs:

Xiaobo Dong

Kerri Hickenbottom

1:30 PM (Blackbird Studio A, Wednesday)

Abstract ID- 335

Ultraselective ammonium recovery from wastewater: membrane processes drive materials design

Oded Nir (Ben-Gurion University of the Negev) *

2:00 PM (Blackbird Studio A, Wednesday)

Abstract ID- 311

Green(er) Solvents for Upcycling Waste PVC into Membranes

ATTA UR RAZZAQ (The University of Alabama) *, Milad Esfahani (The University of Alabama)

2:30 PM (Blackbird Studio A, Wednesday)

Abstract ID- 433

Methanol separation from poly(lactide) depolymerization products

Suer Kurklu Kocaoglu (The University of Texas at Austin) *, Sabah Najjar (The University of Texas at Austin), Carson Womack (The University of Texas at Austin), Louise Kuehster (The University of Texas at Austin), Hamidreza Mahdavi (Monash University), Nirvana Almada (The University of Texas at Austin), Laila Halim (Monash University), Michael Batten (CSIRO), Leonie van't Hag (Monash University), Zongli Xie (CSIRO), Zachariah A. Page (The University of Texas at Austin), Matthew R. Hill (Monash University), Nathaniel A. Lynd (The University of Texas at Austin), Benny D. Freeman (The University of Texas at Austin)

Oral 44- Module and Systems Modeling

1:30-3:30 PM: Blackbird Studio B

Session chairs:

Yang Han

Harekrushna Behera

1:30 PM (Blackbird Studio B, Wednesday)

Abstract ID- 99

Deciphering the Impact of Transmembrane Pressure on Reverse Osmosis Composite Membrane Compaction and Performance

Jishan Wu (Rice University) *, Menachem Elimelech (Rice University), Ying Li (University of Wisconsin-Madison), Eric Hoek (UCLA)

2:00 PM (Blackbird Studio B, Wednesday)

Abstract ID- 297

Validation of high-flux plate-and-frame membrane module design for CO_2 capture: a CFD Approach

Cheick Dosso (Carnegie Mellon University) *, Hector Pedrozo (Carnegie Mellon University), Thien Tran (National Energy Technology Laboratory), Lingxiang Zhu (National Energy Technology Laboratory), Victor Kusuma (National Energy Technology Laboratory), David Hopkinson (National Energy Technology Laboratory), Lorenz Biegler (Carnegie Mellon University), Grigorios Panagakos (Carnegie Mellon University)

2:30 PM (Blackbird Studio B, Wednesday)

Abstract ID- 255

Parametric Optimization of Layer-by-Layer Coated Ion Exchange Membranes: Insights from Computational Modeling

Omar Elshawarby (KU Leuven) *, Xing Yang (KU Leuven)

3:00 PM (Blackbird Studio B, Wednesday)

Abstract ID- 269

The influence of spacer arm identity on salt transport in charged nanofiltration membranes

Laurianne Estrada (University of Notre Dame) *, Jonathan Ouimet (University of California, Santa Barbara), William Phillip (University of Notre Dame)

Oral 45- Electrochemical Separations II

1:30-3:30 PM: Southern Ground AB

Session chairs:

David Ladner

Dan Miller

1:30 PM (Southern Ground AB, Wednesday)

Abstract ID- 195

Thiol-Ene Photopolymerization for Highly Selective Anion Exchange Membranes with High Charge Concentration

Jocelyn Riet (University of Toronto) *, Sara Abu-Obaid (Columbia University), Jay Werber (University of Toronto)

2:00 PM (Southern Ground AB, Wednesday)

Abstract ID- 225

Flow-through permeability apparatus for automated high-throughput analysis of complex solute mixtures

Aubrey Quigley (The University of Texas at Austin) *, Kris Gleason (The University of Texas at Austin), Nathaniel Lynd (The University of Texas at Austin), Benny Freeman (The University of Texas at Austin)

2:30 PM (Southern Ground AB, Wednesday)

Abstract ID- 79

Pillared Laminar Phyllosilicate Membranes with Tunable Ion Sieving

Yining Liu (University of Chicago) *, Yuqin Wang (University of Chicago), Bratin Sengupta (Argonne National Laboratory), Alex Martinson (Argonne National Laboratory), Jeffrey Elam (Argonne National Laboratory), Seth Darling (Argonne National Laboratory)

Oral 46- Desalination and Osmotic Processes II

1:30-3:30 PM: Ocean Way

Session chairs:

Thien Tran

Lakshmeesha Upadhyaya

1:30 PM (Ocean Way, Wednesday)

Abstract ID- 252

Concentration of organic solvents by osmotically assisted reverse osmosis

Mengyang Hu (Kobe University) *, Hideto Matsuyama (Kobe University)

2:00 PM (Ocean Way, Wednesday)

Abstract ID- 405

Solution-friction approximation as a robust theoretical framework for modeling low-salt-rejection reverse osmosis

Rayan Alghanayem (Vanderbilt University) *, Rui Chen (Vanderbilt University), Weifan Liu (Vanderbilt University), Shihong Lin (Vanderbilt University)

2:30 PM (Ocean Way, Wednesday)

Abstract ID- 52

Enhancing forward osmosis efficiency: unraveling the physicochemical interactions between Layer-by-Layer coated membranes and Pluronic® as a draw solution

Aylin Kinik (Eindhoven University of Technology) *, Marrit van der Wal (Eindhoven University of Technology), Kitty Nijmeijer (Eindhoven University of Technology), Zandrie Borneman (Eindhoven University of Technology)

3:00 PM (Ocean Way, Wednesday)

Abstract ID- 235

Salty no More: Explainable AI-Driven Discovery of Polyamide Desalination Membranes

Subhamoy Mahajan (University of Wisconsin-Madison), Ying Li (University of Wisconsin-Madison) *

Oral 47- Inorganic Membranes

1:30-3:30 PM: Melody Room

Session chairs:

Caleb Funk

Katie Li-Oakey

1:30 PM (Melody Room, Wednesday)

Abstract ID- 381

Ceramic Ion Exchange Membrane Operational Stability in Organic Solvents

Stephanie Candelaria (Membrion) *, Ryan Flores (Membrion), Unnati Rao (Membrion), Emily Rabe (Membrion), Greg Newbloom (Membrion)

2:00 PM (Melody Room, Wednesday)

Abstract ID- 418

Scalable Bottom-Up Synthesis of Nanoporous Hexagonal Boron Nitride (h-BN) for Large-Area Atomically Thin Ceramic Membranes

Andrew Naclerio (Vanderbilt University), Peifu Cheng (Vanderbilt University) *, Saban Hus (Oak Ridge National Laboratory), Trey Diulius (National Institute of Standards and Technology), Marti Checa (Oak Ridge National Laboratory), Ivan Vlassiouk (Oak Ridge National Laboratory), William Fissell (Vanderbilt University Medical Center), Matthew Coupin (The University of Texas at Austin), Jamie Warner (The University of Texas at Austin), Liam Collins (Oak Ridge National Laboratory), Andrei Kolmakov (National Institute of Standards and Technology), An-Ping Li (Oak Ridge National Laboratory), Piran Kidambi (Vanderbilt University)

2:30 PM (Melody Room, Wednesday)

Abstract ID- 260

Investigation of Hydrogen Permeation and Material Characterization of Pd-Ag-Y Membrane

Omid Jazani (Clarkson University) *, Simona Liguori (Clarkson University)

Polymers of Intrinsic Microporosity: Promising materials for water desalination

Mahmoud Abdulhamid (King Fahd University of Petroleum and Minerals) *

Hexa-fluorinated porous membranes were synthesized using 4,4-(hexafluoroisopropylidene) diphthalic anhydride (6FDA) and 2,3,5,6-tetramethyl-p-phenylenediamine (TMPD) through a polycondensation reaction at 200 °C, followed by a nonsolvent-induced phase inversion process. These 6FDA-TMPD membranes were specifically designed to address the challenge of desalinating highly saline water (70,000 mg L−1) using membrane distillation (MD) technology, an advanced separation process that operates on the principle of water vapor transport through a porous hydrophobic membrane. The study highlights the critical role of intrinsic porosity in these membranes, which not only enhances their performance but eliminates the need for additional porogen chemicals commonly used in other membrane fabrication processes. This intrinsic porosity provides the membranes with a significant advantage over conventional and commercially available membranes by delivering superior desalination efficiency and stability under harsh operating conditions.

To optimize the performance of the membranes, key fabrication parameters were systematically investigated. Factors such as the coagulation bath composition, dope solution concentration, and membrane thickness were fine-tuned to achieve the desired properties. A standout feature of the developed membranes is their remarkably high porosity, attributed to the large BET surface area of 550 m^2 g−1, which facilitates efficient water vapor transport. The use of an ethanol-based coagulation bath further enhanced the membrane morphology, resulting in a spongy-like structure that proved highly effective in the MD process. This structural optimization enabled the membranes to achieve an impressive water vapor flux of approximately 18.20 kg m−2 h−1 and a stable salt rejection rate exceeding 99.96%. This study paves the way for developing advanced polymer materials for effective water desalination applications.

Sharklet® imprinted patterns on thin-film composite membranes for scaling resistance in water desalination

Jacalyn Morgan (University of Colorado Boulder) *; Dhairya Agnihotry (University of Colorado Boulder); Caleb Song (University of Colorado Boulder); Keiran Fung (University of Colorado Boulder); John Pellegrino (University of Colorado Boulder)

This study investigates the potential of incorporating regular surface patterns on commercially produced polyamide thin-film composite (TFC) reverse osmosis (RO) and nanofiltration (NF) exemplar membranes to enhance scaling resistance. Using embossing, we successfully patterned TFC membranes with Sharklet® surface features at micron-sized patterns, intending to enhance scaling resistance. Results showed that this micron-sized pattern increased surface area for the NF membrane and overall permeance was either increased or left unchanged. Surface roughness was highly retained in the pattern features for CR100 and UHPRO RO membranes, indicating potential relationships between pattern feature size, membrane compression, and mass transfer properties. Performance testing using NaCl solution flux and salt rejection confirmed the retention of the membranes' desalination integrity post-patterning. Closed-circuit desalination tests with supersaturated solutions producing calcite and gypsum precipitates demonstrated improved scaling resistance in patterned membranes compared to unpatterned counterparts. SEM imaging further highlighted the clarity of Sharklet® features, supporting their potential to mitigate scale adhesion. These findings suggest that micro-patterned membranes maintain their desalination functionality while exhibiting promising scaling resistance, making them a viable approach for enhancing the longevity and efficiency of TFC RO membranes.

Synchronous electrosynthesis and separation of ammonia and chlorine from waste brinestreams via a stacked membrane-free electrolyzer

Wen Zhang (New Jersey Institute of Technology) *

Electrosynthesis offers a promising route for decarbonizing the chemical industry by generating valuable chemicals under ambient conditions. We introduce a membrane-free flow electrolyzer designed for paired electrocatalytic upcycling of nitrate (NO_3^-) and chloride (Cl^-) to produce ammonia (NH_3) and chlorine (Cl_2) gases, utilizing waste streams as alternative electrolytes. This electrolyzer integrates electrosynthesis and gaseous-product separation, minimizing redox reactions between NH_3 and Cl_2, thus reducing product loss.

The system features a three-stacked-module electrolyzer setup capable of efficiently processing reverse osmosis retentate waste. This approach achieved high concentrations of $(NH_4)_2SO_4$ (83.8 mM) and NaClO (243.4 mM) at an electrical cost of 7.1 kWh per kilogram of solid products, while residual pollutants (NH_3/NH_4^+, NO_2^-, and Cl_2/ClO^-) remained within wastewater discharge regulations.

The electrolyzer operates across three main stages: (1) electrochemical conversion of NO_3^- and Cl^- ions to NH_3 and Cl_2, (2) vaporization of NH_3 and Cl_2 at respective basic and acidic interfaces, and (3) interfacial extraction at the electrode surfaces. Pairing nitrate reduction with chloride oxidation eliminated undesired by-products (e.g., H_2 and O_2). A specially designed gas extraction electrode concurrently coupled electrosynthesis and product separation, allowing simultaneous NH_3 and Cl_2 generation and separation at the interface.
The scalability of this approach was demonstrated with a stacked system using a 300 cm^2 electrode area and real reverse osmosis retentate waste streams. Our findings underscore the potential for coupling NH_3 and Cl_2 production and separation within a single, streamlined electrolyzer design. This research paves the way for innovative, sustainable chemical production systems.

Water-Pore Flow Permeation through Multivalent Artificial Water Channels

Mihail BARBOIU (Institut Europeen des Membranes) *

Aquaporins (AQPs) are natural proteins able to selectively transport water across cell membranes. Heterogeneous H-bonding of water with inner wall of the pores of AQPs is of maximal importance regarding the optimal stabilization of water clusters within channels leading to selective water transport against ions. To get deeper insight into the water permeation mechanisms, simpler artificial water channels (AWCs) have been developed. Several H -bonding motifs (i.e. imidazole, polyhydroxy, etc.) have been reported as distinct and efficient for water-cluster stabilization within AWCs. Herein we describe several membrane pore configurations able to stabilize, like in AQPs water molecules via different H-bonding groups. The crystal structures reveal that stable water superstructures are formed in the solid state, accommodating water clusters or water-wires. Molecular simulation confirmed that AWC-channels can generate stable supramolecular porous sponges, featuring multivalent water H-bonding units that serve as water-cluster relays within the channel. To the best of our knowledge, these works are a rare biomimetic example of the importance of water-cluster stabilization via multivalent H-bonding and toward selective transport through water channels and further translated for the fabrication of efficient biomimetic membranes for desalination.

1. D.D. Su, M. Barboiu Coord. Chem. Rev. 2024, 515, 215973.
2. L.-B. Huang, M. Di Vincenzo, Y. Li, M. Barboiu,. Chem. Eur. J., 2021, 27, 2224-2239.
3. I.-M. Andrei, W. Chen, M. Baaden, S. Vincent, M. Barboiu, J. Am. Chem. Soc., 2023, 145, 40, 21904–21914
4. M. Di Vincenzo, A. Tiraferri, V.-E. Musteata, S. Chisca, M. Deleanu, F. Ricceri, D. Cot, S. P. Nunes, M. Barboiu, Proc. Natl. Acad. Sci., 2021, 118(37), e2022200118.
5. M. Di Vincenzo, A. Tiraferri, V.-E. Musteata, S.Chisca, R. Sougrat, L.-B. Huang, S. P. Nunes, M. Barboiu, Nanotechnol. 2021, 16, 190-196.

On the role of the porous support membrane in seawater reverse osmosis membrane synthesis, properties and performance

Eric Hoek (UCLA) *

Reverse osmosis (RO) membranes most often comprise "thin film composite" (TFC) structures, where an ultra-thin, highly-crosslinked microporous coating film is formed on top of an asymmetric meso-to-macro-porous support layer. The selective layer can be synthesized by interfacial polymerization of polymers such as 1,3-diaminobenzene (or "meta-phenylenediamine," MPD) and 1,3,5-benzenetricarbonyl trichloride (or "trimesoyl chloride," TMC). The separation performance of these polyamide (PA) coating films is governed by their mass/thickness, morphology, crosslinking degree (XD), and charge density; however, the porous support membrane skin layer pore morphology also contributes to TFC membrane separation performance by altering the effective permeation path length through the composite structure. The combination of tuning the support membrane pore morphology and the coating film chemistry and structure gives rise to the wide tunability of TFC-RO membrane properties to achieve performances ranging from nanofiltration, "ultra-low pressure" RO membranes for water recycling, "low-pressure" RO membranes for brackish water desalination, and "high-pressure" RO membranes for seawater desalination.

Herein, we hypothesize that the support membrane skin-layer pore size and porosity, and whole-body porosity govern the amount of MPD uptake and release, and hence, governs the MPD:TMC ratio in the reaction zone, which produces PA films of different mass/thickness, XD, and separation performance. We vary PSU support membrane pore morphology by varying the PSU concentration and the temperature of the coagulation bath. We characterize PSU support membrane and composite PA-PSU membrane SWRO separation performance and physical-chemical properties. Finally, we demonstrate that by varying the PSU support pore morphology, one can fine-tune the PA film mass/thickness, XD and separation performance to achieve SWRO performance on par with commercial SWRO membranes.

Texture-Driven Anti-Biofilm Efficacy of Laser-Induced Graphene: A Promising Solution to Control Membrane Biofouling

Kritika Jashrapuria (Indian Institute of Technology Bombay) *; Swatantra Pratap Singh (Indian Institute of Technology bombay)

Bacterial biofilms pose significant challenges in membrane-based industries, including food, water, and wastewater treatment. An effective strategy to mitigate bacterial biofouling involves altering surface texture to inhibit the attachment, colonization, and biofilm formation of bacterial cells. Nanostructured textured surfaces have shown remarkable efficacy in reducing biofouling by preventing single-cell adhesion and biofilm development. Thus, in this study, we investigated the role of Laser-Induced Graphene (LIG) texture in biofilm inhibition and found that rough LIG surface nano-textures correlate with enhanced biofilm prevention. We examined the interactions of LIG nano-fibers with two different bacterial species: Escherichia coli and the Staphylococcus epidermidis. Rough LIG nano-textures effectively inhibited bacterial attachment and colonization by limiting the surface area availability for their settlement. The LIG nano-fibers disrupted bacterial proliferation within their networks, inducing morphological changes and reducing E. coli cell size by ~40% due to the General Stress Response (GRS). LIG surfaces were integrated into electro-conductive composite membranes and filters to further explore their anti-biofouling potential for water treatment. These composite membranes exhibited superior filtration performance, achieving a high-water flux of ~300 L m^{-2} h^{-1} and more than 50% reduction in live biomass compared to pristine membranes. These findings highlight LIG-based surfaces and membranes as promising anti-biofouling solutions for diverse membrane-based applications.

Keywords: Biofilms, Texture, Anti-biofouling, Laser-Induced Graphene

Synthesis of Recyclable Microporous Polyimide-Based Membranes for Membrane Distillation

Toheeb Obidara (Northeastern University) *; Turki Baroud (King Fahd University of Petroleum Resources)

Membrane distillation (MD) is a promising technology for treating seawater and wastewater. However, MD membranes require significant improvements to ensure excellent separation efficiency. Herein, novel recyclable polyimide-based membranes were fabricated with tailored morphologies through a modified vapor-induced phase separation technique using hexa-fluorinated fluoroisopropylidene anhydride (6FDA) and tetra-p-methyldiamine (TMPD), referred to as 6FDA-TMPD polymer. This process involves controlled exposure of polymer solutions to vapor, to precisely control the membrane porous structure and surface characteristics without chemical modifications. Notably, controlling the vapor type (ethanol instead of water) and exposure duration transforms the membrane's surface structure from a spherical, nodule-like to a highly porous and open interconnected structure, resulting in a significant increase in hydrophobicity, yielding a water contact angle of 141.6o. Moreover, the flux increased by 40% from 14.59 to 20.85 kg m–2 h–1, while maintaining excellent salt rejection of ~ 99.70% when treating high-salinity solutions (70,000 ppm) in air-gap membrane distillation. These findings provide valuable insights into the synthesis and optimization of high-performance 6FDA-TMPD membranes and propose a sustainable and facile alternative to conventional fabrication methods.

Molecular Simulations Reveal Gas Transport Mechanisms in Polyamide Membranes

Ruoyu Wang (Rice University) *; Jianhao Qian (Rice University); Menachem Elimelech (Rice University)

Membrane-based gas separation technologies have emerged as a promising alternative to conventional energy-intensive processes such as cryogenic distillation and chemical absorption, offering lower energy consumption, scalability, and reduced carbon footprints. Despite significant advancements, the trade-off between gas permeability and selectivity, governed by the intrinsic structure of polymeric membranes, remains a critical limitation, capping membrane performance at the so-called upper-bound. Addressing this challenge requires a deeper understanding of gas transport mechanisms and structure-performance relationships.

In this study, we employed non-equilibrium molecular dynamics (NEMD) simulations to investigate gas transport through polyamide (PA) membranes and compared the results to transport through water films. Our findings reveal a pronounced molecular sieving effect in PA membranes, with smaller gases such as helium and hydrogen permeating the membrane, while larger gases, including carbon dioxide and methane, fail to traverse the membrane within the simulation timescale. We demonstrate that gas transport in PA membranes deviates from the classical solution-diffusion (SD) model, with stable pore structures resulting in non-uniform gas distributions and directional trajectories. In contrast, the dynamic hydrogen-bond network of water films facilitates uniform gas distribution and random-walk behavior.

These insights highlight the unique gas transport characteristics of PA membranes and suggest that their transport mechanisms, particularly for small gases, may be better described by alternative models, such as Knudsen diffusion. Our work underscores the importance of advancing fundamental knowledge to overcome the permeability-selectivity trade-off and guide the development of next-generation membranes for high-performance gas separations.

Experimental Validation of the Centrifugal Reverse Osmosis (CRO) Process

William Krantz (University of Colorado at Boulder) *; Lee Nuang Sim (Nanyang Technological University); Tzyy Haur Chong (Nanyang Technological University)

The cost of desalination can be reduced if RO were operated at local thermodynamic equilibrium across the membranes. RO at pressures only differentially removed from equilibrium is possible via the Centrifugal Reverse Osmosis (CRO) process that rotates the membrane module to progressively increase the retentate-side pressure. This presentation discusses the first bench-scale experimental validation of CRO, whose configuration is shown in Fig. 1. Diametrically opposed, counterbalancing, rectangular membrane modules with flat sheet RO membranes are attached to a central rotating shaft. These experiments were operated at 300 rpm, 4 bar maximum pressure, and 2 g/L feed concentration. The water recovery Y was changed via input bypass and retentate flow-control valves. The open circles in Fig. 2 show the measured SECgross (kWh/m3) versus Y for CRO that decreases with increasing Y because of concentration polarization owing to the decreased safety factor. An idealized model that ignores all irreversibilities predicts the SECgross shown by the solid circle. A surprising observation is the two highest recoveries exceed the idealized model prediction indicating that another mechanism is operative in CRO. The open squares in Fig. 2 show the measured SECgross versus Y for SSRO. CRO reduces the SECgross relative to SSRO by as much as 26%. An idealized SSRO model predicts the SECgross shown by the solid square. All measured SECgross values for SSRO are larger than the predicted idealized performance implying that the mechanism mitigating concentration polarization in CRO is not operative in SSRO. This mitigating mechanism also significantly reduces membrane fouling as shown in Fig. 3. For a foulant of 0.2 g/L sodium alginate and 2 g/L sodium chloride, SSRO shows a progressive flux decline; in contrast, CRO maintains a constant flux within experimental error over 12 hours. The mitigating mechanism operative in a CRO module will be explained in this presentation.

The physical basis for solvent flow in organic solvent nanofiltration

Hanqing Fan (Rice University) *; Jinlong He (University of Wisconsin-Madiso); Dariush M. Heiranian (North Carolina State University); Weiyi Pan (Rice University); Ying Li (University of Wisconsin-Madison); Menachem Elimelech (Rice University)

Organic solvent nanofiltration (OSN) is an emerging membrane technology that could transform chemical separations across a wide array of critical industries, including pharmaceuticals, petrochemicals, and food processing. Despite its potential to significantly enhance efficiency and sustainability in these applications, there remains a lack of fundamental understanding of the mechanisms governing solvent transport in OSN membranes. In this study, we employ an extended Flory-Rehner theory, nonequilibrium molecular dynamics simulations, and organic solvent transport experiments to elucidate the principles of solvent transport in OSN membranes. Our study reveals that solvent transport is driven by a pressure gradient, not a solution-diffusion mechanism. Solvent molecules migrate in clusters through a network of interconnected pathways within membrane pores. We demonstrate that solvent permeance depends on the affinity between solvent and membrane material, which impacts the pore structure of membranes. In addition, binary solvent mixture permeates through membrane pores as a unified fluid, which further confirms the transport mechanism of viscous flow. These insights provide a comprehensive understanding of the interplay between solvent-membrane interactions and transport behavior. This work lays the scientific groundwork for the rational design and optimization of next-generation OSN membranes.

Enhancing Li+/Mg2+ Separation Properties by Optimizing Positive Charges of NF Membrane Surfaces

Sagnik Das (University at Buffalo) *; Haiqing Lin (University at Buffalo); Erda Deng (University at Buffalo); Kai Chen (University at Buffalo); Jada Mowatt (University at Buffalo)

Lithium is vital in energy storage applications, and its recovery from salt brines (containing mainly Mg2+) using membranes has emerged as an important industrial process. However, commercial polyamide-based nanofiltration (NF) membranes often show low Li+/Mg2+ separation factor (SFLi/Mg) due to their similar hydration radii. Herein, we show that the SFLi/Mg can be dramatically enhanced by optimizing the membrane surface positive charges, and it reaches maximal values at pH values slightly below the isoelectric points (IEPs). Specifically, NF270 membrane was surface-grafted by 2-(methacryloyloxy)ethyltrimethylammonium chloride (META) or polyethylenimine (PEI) assisted by dopamine, forming a thin, stable, positively charged layer (20 – 40 nm) on the membrane surface. Dopamine, in the presence of oxygen, self-polymerizes to form an insoluble polydopamine (PDA) layer, and its amine groups react with the methacrylate groups in META through Michael's addition and PEI via Schiff base and Michael's addition reaction. The effect of the surface modification and solution pH values on the surface zeta potential (ZP) and SFLi/Mg properties is thoroughly investigated. Solution pH plays a significant role in determining the SFLi/Mg values. Mixed salt SFLi/Mg values are slightly higher than the single-salt values because the unbalanced Cl- ions pull smaller Li+ ions in the permeate side, further reducing their rejection. For instance, co-depositing META with PDA increases the ZP from 9.2 to 16 mV and increases SFLi/Mg from 34 to 80 at pH = 4 when treated by a salt solution (2000 ppm) with a Mg2+:Li+ mass ratio of 5.0, surpassing the state-of-the-art NF membranes. Further reducing the pH to 1.0 leads to a decrease in SFLi/Mg to 3.2 due to a strong repelling force towards both ions. The membrane exhibits stable separation properties over 30-h with mixed salt solutions, showcasing its potential for practical applications.

Decoupling the Influence of Sorption and Diffusion on Membrane Permselectivity in Pervaporation

Sk Md Ali Zaker Shawon (Vanderbilt University) *

Membrane-based pervaporation (PV) is a well-established technique for separating azeotropes and close-boiling mixtures. Permselectivity, a crucial performance metric, arises from the combined effects of sorption and diffusion selectivity, typically measured in its entirety from PV experiments. This study, deconvolutes the impact of sorption and diffusion selectivity from overall permselectivity, providing a deeper understanding of membrane performance. We explore the performance of crosslinked polyvinyl alcohol (PVA) membranes in ethanol dehydration using the PV process. Sorption and PV experiments were conducted separately, analyzing swelling, water and ethanol uptake, effective sorption selectivity, and permselectivity across various feed concentrations. The results revealed that swelling is directly proportional to water concentration in the feed, and effective water sorption selectivity is significantly influenced by feed water concentrations. The experimental data indicates that water flux is driven by self-plasticization and is proportional to feed water concentration, whereas ethanol flux is influenced by crossed plasticization, which is affected by the water's plasticization and transport within the membrane. These findings highlight the importance of understanding sorption and diffusion selectivity independently for efficient separation, offering valuable insights for troubleshooting membranes with poor overall selectivity.

Fabrication of Photocatalytic TiO2-f-MWCNTs Biodegradable Polylactic acid Mixed Matrix Membrane for Enhanced Anti-Fouling Properties

Farah Abuhantash (Khalifa University) *

Amidst the urgent need to address environmental imperatives, and significance as industrial water usage is anticipated to undergo a staggering 400% increase by the year 2050, membrane technology is an efficient, economical and environmental solution. However generally, membrane's efficacy hinges on resolving the challenge of rapid fouling. This study aims to develop a photocatalytic mixed matrix biodegradable based membrane using TiO2 and multi-walled carbon nanotubes (MWCNTs) as nanoparticles (NPs) and Polylactic acid (PLA) as the base polymer with enhanced self-cleaning properties.

Photocatalytic TiO2 functionalized MWCNTs NPs were initially synthesized and through phase inversion casting method, membranes with varying nanoparticles concentrations were fabricated. The nanoparticles are characterized by XRD, SEM, EDS and photocatalytic activity, while the composite membranes are tested for water and oil emulsion permeation performance, rejection and long-term anti-fouling with and without photocatalysis at different nanoparticles loading to find the optimal and further characterized by SEM, EDS, Zeta potential, porosity and contact angle.

A range of 0-2wt% NPs loading were tested, the 2 wt% TCNTs loading to the PLA matrix boosted water permeability by 89% compared to pristine, with up to 99.4% oil rejection. Furthermore, a 5-cycle study using UV radiation as a cleaning technique revealed the cyclic permeate total organic carbon (TOC) levels of the PLA/2-TCNTs to stay within EPA TOC standards even after 5 cycles of filtration, and 100% FRR by the 4th cycle, while pristine nearly reached the TOC limit on the second cycle and exceeded by the fifth cycle. The prepared photocatalytic membranes have a self-cleaning property for oily wastewater and enhanced membrane anti-fouling tendency with reduced environmental impact by the use of recyclable polymers, such as PLA.

Combined Nanofiltration and Diafiltration for Isolation of Rare-earth Ions

Jun Wang (University of Notre Dame) *; Dong Ding (University of Notre Dame); Laurianne Lair (University of Notre Dame); Andriy Yaroshchuk (Polytechnic University of Catalonia); William Phillip (University of Notre Dame); Merlin Bruening (University of Notre Dame)

Rare-earth elements (REEs) are essential for renewable energy technologies, but their extraction and purification are difficult due to the presence of competing ions in recycled and natural sources. Conventional separation methods often require large amounts of chemicals and produce significant waste, posing economic and environmental challenges. This study demonstrates a combined nanofiltration (NF) and diafiltration process for efficiently purifying REEs, focusing on La^{3+} separation from mixed-ion solutions.

The combined NF and diafiltration approach achieves a final La^{3+} molar purity of 99%, starting from equimolar mixtures with Na^{+}, demonstrating exceptional performance for selective REE concentration. Highly negative rejections of monovalent ions, such as Na^+ and H^+, significantly enhance the purification step (In negative rejection, the concentration of a specific ion is higher in the permeate than in the feed.). In diafiltration, proton concentrations decrease 3–4 orders of magnitude within 2.4 cell volumes of filtration, highlighting the system's capability to remove acid efficiently while retaining REEs.

We model and optimize this process using a rotating membrane cell to achieve a controlled, uniform boundary layer, which facilitates estimation of ion permeances at various feed compositions. Using a solution-diffusion-electromigration model, we analyze trends in ion rejection as a function of feed composition, flux, and boundary layer properties.

These findings underscore the potential of membrane-based separations as scalable and environmentally sustainable solutions for REE purification and recovery. Future studies will explore broader implementation across REE mixtures, enhanced acid removal at lower pH, and compatibility with ligand-assisted separations to refine selectivity.

Amphiphilic Polyelectrolyte Complex Bilayer Membranes

Nikan Mokarinezhad (Tufts University) *; Luca Mazzaferro (Tufts University); Usama Sheikh (University of Connecticut); Benjamin J. Sundell (ZwitterCo, Inc.); Jeffrey R. McCutcheon (University of Connecticut); Ayse Asatekin (Tufts University)

Membranes are efficient, scalable, and simple to operate. However, traditional membranes often exhibit limited selectivity and propensity to fouling. Their long-term chemical and physical stability is also limited. The ability of functional polymers to self-assemble provides a basis for creating membranes that can perform novel and effective separations. Our group has introduced a scalable method to create such high- performance thin film composite (TFC) membranes with Amphiphilic PolyElectrolyte Complex (APEC) selective layers, which self-assemble to form tight "nanochannels" with tunable selectivity. Amphiphilic PolyElectrolytes (APEs) are random copolymers that combine hydrophobic and charged monomers and are water-insoluble. In this process, cationic and anionic APEs are sequentially deposited over a substrate to create a bilayer TFC membrane. When APEs contact at the interface, they self-assemble into the complex form (APEC) due to interactions between opposite charged functional groups in two APE layers. This creates nano-sized ionic domains, acting as selective channels within a hydrophobic matrix and produces TFC membranes with organics rejection in nanofiltration range coupled with high permeability. To date, only one APEC chemistry has been studied. This work explores how the chemical structures of hydrophobic and charged monomers forming the APEs affect the performance and capabilities of membranes. Preliminary results indicate that these membranes exhibit high water permeance and stable tight pore sizes, with high small organics and comparatively lower monovalent salt rejections. Having low salts and high neutral solutes rejections is beneficial for many applications. A simple and scalable method of synthesis and membrane manufacturing was used with diverse potential chemical functionalities which offers versatility and potential for creating membranes with targeted selectivity, promising for water treatment applications and complex separations.

Selective Ion Transport through Hydrated Micropores in Polymer Membranes

ANQI WANG (King Abdullah University of Science and Technology (KAUST) *

With increasing reliance on renewable but intermittent energy sources like solar and wind, electrochemical technologies are essential for on-site energy storage and integrating low-carbon power into the grid. Aqueous organic redox flow batteries (RFBs) offer great potential for grid-scale, long-duration energy storage, but their development is limited by the need for ion-selective polymer membranes. Here, I will discuss our recent efforts in the development of microporous membranes designed for precise ion transport in RFB applications. Through modular synthesis of microporous polymers, we introduce ion-coordinating functionalities to enhance rapid ion transport, while optimizing pore geometry and channel topology in hydrated membranes for efficient size-sieving selectivity. These membranes demonstrate faster ion conduction and orders-of-magnitude higher ion selectivity than commercial alternatives, resolving the trade-off challenges typical of conventional membranes. When paired with energy-dense organic redox couples, these membranes enable higher energy efficiency and significantly lower capacity decay in RFB systems.

References: Wang, A., Breakwell, C., Foglia, F. et al. Selective ion transport through hydrated micropores in polymer membranes. Nature 635, 353–358 (2024). https://doi.org/10.1038/s41586-024-08140-2

Stable Two-Dimensional Siloxene Membranes for Molecular Separations

Dinesh Mahalingam (University of Bath) *

Membranes fabricated using two dimensional (2D) materials have shown immense promise in water purification. However, their use remains limited owing to the membrane instabilities in aqueous environments: Graphene oxide and boron nitride membranes undergo swelling, resulting in structural disintegration and membrane failure. Here, we report for the first time highly stable 2D siloxene membranes obtained from a scalable exfoliation process and requiring no additional cross-linking or functionalisation to retain their stability. The 2D siloxene selective layers, on a porous nylon support, have a high mechanical integrity even at 6 bar operating pressure, exhibiting a pure water permeance of 9 L m-2 h-1 bar-1 and high rejection of organic dyes over 697 g mol-1 molecular weight. This performance is comparable to the commercial polymeric membranes used for nanofiltration. The stability of the membranes was tested under crossflow conditions by repeatedly drying and soaking them in water over the course of a month, applying different pressure regimes and repeated dye rejection tests. This stability of siloxene membrane is due to the strong hydrogen bonding network between its hydroxyl groups and nylon as revealed by FT-IR and XPS studies. The stability of siloxene membranes and their scalable manufacturing method mean they are a realistic candidate to overcome the limitations of existing 2D material-based membranes which have, so far, limited their applicability to address performance challenges. Furthermore, the use of 2D siloxene material opens new opportunities in membrane development as it allows leveraging well-established silicon chemistry, hitherto little used in membrane science and technology.

Alternating Potential Operation of Electro-Conductive Membranes: A Dual Approach to Mitigating Biofouling and Scaling in Advanced Wastewater Treatment

Avner Ronen (Ben Gurion University) *; Heyang Yuan (Temple University); Wen Zhang (NJIT); Siyao Qi (Ben Gurion University); Roy Bernstein (Ben Gurion University); Zhang Cheng (Temple University)

Electro-conductive membranes (ECMs) have emerged as a promising low-fouling technology for advanced tertiary wastewater treatment. When utilized as cathodes, ECMs demonstrate significant resistance to biofouling. However, elevated pH levels near the cathode can lead to calcium and magnesium-induced inorganic fouling, thereby increasing transmembrane pressure (TMP). Despite these advantages, there remains a critical gap in understanding how to simultaneously mitigate both biofouling and scaling under real operational conditions, particularly when treating complex secondary effluents.

This study explores an operational strategy using alternating potentials to simultaneously address biofouling and scaling while producing high-quality effluent. An electrochemical filtration cell equipped with an ion exchange membrane and a carbon nanotube (CNT)-based ECM was employed. Alternating cathodic and anodic potentials at 5-second intervals were tested for their ability to mitigate scaling.

Under optimized conditions, a 48-hour experiment using real secondary effluents revealed that alternating potential operation maintained a significantly lower TMP (140 mbar) compared to open-circuit potential (OCP) conditions (~370 mbar). Scanning electron microscopy confirmed the absence of scaling on the ECM after alternating potential operation, whereas a biofouling layer developed under OCP conditions. Confocal laser scanning microscopy further demonstrated a reduced volume of live bacteria after alternating potential operation. This reduction was attributed to the generation of reactive oxygen species and hydrogen peroxide, which facilitated bacterial detachment. Additionally, the short anodic phases under alternating potential conditions appeared to minimize CNT oxidation, potentially extending the ECM's lifespan. Finally, the ECM-based filtration produced superior effluent quality compared to conventional tertiary treatment methods such as chlorination-coagulation and sand filtration.

Rinse-Free Deposition of Molecular Layer-by-Layer (mLbL) Polyamide Reverse Osmosis Membranes

Samarpan Deb Majumder (George Washington University)*; Christopher Stafford (National Institute of Standards and Technology); Xitong Liu (George Washington University)

The molecular layer-by-layer (mLbL) deposition technique enables the fabrication of polyamide-based reverse osmosis membranes with low surface roughness and tunable membrane thickness, but scale-up of the process is challenged by the necessity for rinsing away unreacted monomers after each deposition step. By sensible tuning of monomer concentrations during deposition, we can eliminate the rinsing steps while still producing polyamide membranes with comparable thickness and surface properties. This approach markedly shortens fabrication time and improves reagent efficiency. By measuring the growth rate of the polyamide membrane as a function of deposition cycles and monomer concentration, we demonstrate that we can deliver a more targeted quantity of monomer to the growing film surface while maintaining a linear growth rate profile. Fourier Transform Infrared (FTIR) spectroscopy shows a negligible amount of the unreacted (free) carboxylic acids across the monomer concentrations, indicating that the crosslink density of the membrane remains high. Desalination tests reveal that mLbL membranes produced at the lowest monomer concentration (membrane thickness of ≈20 nm) achieved nearly 99 % salt rejection within 3h, while maintaining a flux of $0.1 Lm^{-2}h^{-1}bar^{-1}$. This modified mLbL approach significantly reduces membrane fabrication time without compromising performance, offering a path towards scalable production of polyamide-based membranes with controlled thickness and low roughness while prioritizing both time and resource efficiency.

Design and Characterization of Inter-crosslinked Nanocomposite Membranes Exhibiting Stable Performance in Organic Solvent Separations

Lucas Condes (University of Oklahoma)*; Michele Galizia (University of Oklahoma)

The authors chose not to publish the abstract content.

Reducing Physical Aging of Microporous Polymer Membranes through Blending with Porous Polymer Networks

Lucas Condes (University of Oklahoma)*; Matthew Webb (University of Oklahoma); Tran Le (University of Oklahoma); Will Box (Air Products); Cara Doherty (CSIRO); Aditi Gali (University of Oklahoma); Leoncio Garrido (ICTP-CSIC); Jing Deng (ALTR, FLTR Inc.); Laura Matesanz (ICTP-CSIC); Angel Lozano (ICTP-CSIC); Cristina Alvarez (ICTP-CSIC); Marco Nardelli (University of North Texas); Alberto Striolo (University of Oklahoma); Anita Hill (CSIRO); Michele Galizia (University of Oklahoma)

The authors chose not to publish the abstract content.

Fabrication and Scale-Up of Porous Polybenzimidazole (PBI) Supports for Gas Separation Composite Membranes

Lingxiang Zhu (National Energy Technology Laboratory)*; Fangming Xiang (National Energy Technology Laboratory); Thien Tran (National Energy Technology Laboratory); Victor Kusuma (National Energy Technology Laboratory); David Hopkinson (National Energy Technology Laboratory)

Industrial gas separation often uses thin film composite (TFC) membranes comprising a porous support overlaid with a single-/multi-layer gas-selective thin film. An optimal porous support should possess high surface porosity to minimize gas transport resistance and nano-sized pores to ease pore penetration occurring during the thin film coating process. Good chemical and thermal stabilities are essential to withstand the aggressive solvents and heat required for the thin film coating and curing. However, few porous membranes satisfy all these requirements. This study presents a scalable membrane formation method of making highly porous polybenzimidazole (PBI) supports via non-solvent induced phase separation. Fabrication conditions, including dope compositions and coagulation bath temperatures, were tuned to improve membrane porosity and create sub-50 nm pores. An optimized PBI support showed surface porosity of 20%, pore diameter of <42 nm, N2 permeance of 300,000 gas permeance unit (GPU), and excellent thermal and chemical stabilities. The PBI support's superior performance for TFC fabrication was validated for both two-layer and multi-layer TFCs in this work. For example, a defect-free polydimethylsiloxane (PDMS) layer cast on the PBI support exhibited extremely high CO2 permeance of 15,000 GPU. Moreover, a multi-layer TFC, formed by applying a PolyActiveTM layer on the PDMS/PBI bilayer, achieved CO2 permeance of 3,100 GPU (with CO2/N2 selectivity of 41), which is 50% more permeable than the best-performing PolyActiveTM TFCs reported for CO2/N2 separation. This presentation will also detail the scale-up fabrication of PBI supports using a custom roll-to-roll membrane casting machine. The combination of excellent support performance and scalable preparation makes the PBI support promising for large-scale TFC production and application.

From Medicine to Water: Dialysis as a Novel Solution for High-Salinity Organic Wastewater Treatment

Yuanmiaoliang Chen (Rice University)*; Menachem Elimelech (Rice University); Zhangxin Wang (Guangdong University of Technology)

High-salinity organic wastewaters pose a major challenge for conventional wastewater treatment processes. In this study, we propose dialysis as an innovative solution to overcome this challenge. Dialysis uses an ultrafiltration (UF) membrane that allows the passage of salts while rejecting the organic substances, operating in a bilateral countercurrent flow mode without the application of hydraulic pressure. Using bench-scale experiments and a model for salt and water transport in leaky membranes, we demonstrate that dialysis can effectively desalinate high-salinity organic wastewaters without diluting the wastewater stream. By comparing the salt/organic selectivity of dialysis and UF using the same membrane, we show that dialysis can effectively fractionate salts and organic substances in high-salinity organic wastewaters. Additionally, we find that, unlike UF, dialysis is almost unaffected by membrane fouling, highlighting its excellent fouling resistance. We conclude by proposing potential high-salinity organic wastewater treatment schemes based on dialysis, paving the way for more sustainable and effective management of challenging wastewaters.

Bridging the Gap Between Pure and Mixed-Gas Performance of Thin-Film Composite Membranes: Advancing Cell Design and Testing for Scalable Module Development

Thien Tran (National Energy Technology Laboratory)*; Victor Kusuma (National Energy Technology Laboratory); David Hopkinson (National Energy Technology Laboratory); Lingxiang Zhu (National Energy Technology Laboratory)

Industrial flue gas, with low CO2 partial pressures (0.05–0.3 bar) and high flow rates, presents a unique challenge for membrane-based post-combustion CO2 capture. Thin-film composite (TFC) membranes, known for their high permeance and efficient use of selective materials, are a promising solution for large-scale deployment. However, the mixed-gas performance of TFC membranes in simulated or real flue gas mixtures remains poorly understood, affecting scale-up efforts for industrial applications. Current lab-scale mixed-gas testing often yields permeance and selectivity values 30% lower than pure-gas values, even at low stage cuts (≤5%). This study addresses the gap between pure- and mixed-gas performance by fabricating high-permeability PolyActive™ TFC membranes (over 3,000 GPU) and testing them in various permeation setups. A novel counter-current permeation cell was developed by combining computational fluid dynamics and 3D printing to optimize flow efficiency and minimize concentration polarization. This custom cell demonstrated a 40% increase in mixed-gas CO2 permeance compared to conventional cells. Additionally, a vacuum permeation testing method further reduced concentration polarization and enhanced mixed-gas CO2 permeance by 41%. These findings challenge current assumptions about mixed-gas performance deficits, emphasizing the importance of advanced cell design and testing methods. This work advances membrane technology toward industrial applications by scaling up from a counter-current mini-cell (~1 cm^2) to a 3D-printed plate-and-frame module (~1,000 cm^2), which was successfully deployed in a field demonstration capturing CO_2 from a major steel producer's blast furnace waste gas. This first-of-its-kind demonstration in the United States underscores the transformative potential of TFC membranes for scalable, sustainable CO2 separation.

Hydrophobic electrically conducting membranes (ECMs) for enhancing ammonia recovery coupled with membrane fouling mitigation

SIYAO QI (Ben-Gurion University)*; Avner Ronen (Ben-Gurion University of the Negev)

Human activities generate significant domestic wastewater with high nitrogen concentrations, posing substantial environmental challenges. Traditional biological nitrogen removal processes convert ammonium to gaseous nitrogen, failing to recover nitrogen as a valuable resource. This inefficiency highlights a gap in wastewater treatment: the need for methods to recover nitrogen in usable forms. Electrically conducting membranes (ECMs) offer an innovative solution by enabling ammonia electrostripping for recovery as liquid fertilizers. Acting as a cathode, the ECM drives water hydrolysis, locally raises pH and converts ammonium to ammonia, which diffuses across the hydrophobic ECM and is stripped by acidic streams.

While pH control is crucial, interfacial mass transport mechanisms—specifically the interaction of pH gradients and applied potentials—are not yet fully elucidated, representing a research gap. Moreover, fouling poses a challenge when treating real wastewater, limiting efficiency and scalability.

This study investigates enhanced ammonia recovery and fouling reduction in an electrochemical membrane stripping system (EMS) utilizing the ECM fabricated by carbon nanotubes. The objectives include optimizing operational factors, analyzing electromigration's role in ammonia recovery through experimental and theoretical methods, and exploring fouling reduction mechanisms using synthetic (SMW) and real (RMW) domestic wastewater.

The EMS achieved 88% ammonia recovery with an ammonium flux of 37 $g/m^2/d$, a 33.5% improvement over conventional membrane stripping systems (CMS) that rely on chemical pH adjustments. Theoretical models confirmed electromigration and elevated ammonia concentrations at the ECM interface as key drivers of mass transfer.

Furthermore, fouling in the CMS declined flux significantly by 34% and 27% for SMW and RMW, respectively, whereas the EMS kept stable flux and clean membranes through hydrogen peroxide production and electrostatic repulsion.

Poly(Terphenyl Fluorenyl Isatin) with Fluorene-Linked Sulfonate Groups and Microblock Structures For Proton Exchange Membranes

Chaoyi Ba (Argonne National Laboratory)*; Shu Xu (Argonne National Laboratory); Christopher Arges (Argonne National Laboratory); Jae Hyung Park (Argonne National Laboratory); Jinkyu Park (Argonne National Laboratory); Meltem Urgun Demirtas (Argonne National Laboratory)

Proton exchange membranes (PEMs) are crucial components in energy storage and conversion devices, such as fuel cells and electrolyzers. Currently, perfluorosulfonic acid (PFSA)-based membranes and ionomers dominate PEM applications due to their high proton conductivity, mechanical robustness, and durability. However, their high costs, relatively high H2 crossover, and environmental concerns have driven the search for PFSA alternatives. Hydrocarbon-based membranes offer potential solutions to these challenges. In this study, we developed a series of non-fluorinated PEMs from poly(terphenyl fluorenyl isatin) with fluorene-pendent bisulfonate groups. These polymers exhibit a microblock structure, consisting of hydrophobic blocks, hydrophilic blocks, and alternating blocks, due to the varying reactivity and solubility of the hydrophobic p-terphenyl and the hydrophilic bisulfonated fluorene monomers. Consequently, the sulfonic acid groups are unevenly distributed along the polymer chains. This microstructure, along with the degree of sulfonation, significantly influences the comprehensive properties of the membranes. The presence of microblocks and hydrogen bonding between amide groups ensures a phase-separated morphology, contributing to the membranes' excellent properties and performance. These include strong mechanical strength (47.1–63.2 MPa), high thermal stability (up to 250°C), low swelling ratio (18–25% at 80°C), and high proton conductivity (136–169 mS/cm in pure water at 80°C). Additionally, the >2x lower hydrogen crossover (0.86 mA/cm2) than Nafion 212 (1.83 mA/cm2) (tested at 80℃, 100% RH) provides the potential ability to go to thinner membranes, which will improve fuel utilization in fuel cell stacks. This work offers valuable insights into the design of robust, high-performance PEMs by systematically analyzing the relationships between membrane structure, properties, and performance.

Advancement of Hybrid Membrane Amine-Absorption Gas Treating for Highly Sour Natural Gas Reserve Development

GARBA YAHAYA (SAUDI ARAMCO)*; FERAS HAMAD (SAUDI ARAMCO); SEBASTIEN DUVAL (SAUDI ARAMCO)

Membrane-amine absorption hybrid – based natural gas sweetening processes have emerged as among the fastest growing technologies, due to their lower capital cost, higher energy savings, smaller size, and more economically viable when compared to the conventional standalone amine absorption process. The technology also offers environmental benefits by significantly mitigating CO2 emissions due to lower energy requirement. This study focuses on the development of aromatic co-polyimide membranes based on 6FDA-Durene/CARDO backbone and investigation of its sour gas separation properties under very harsh sour gas environment (i.e. high H2S content of up to 20 vol% and feed pressure of up to 800 psi for a gas mixture consisting of CO2/H2S/CH4/C2/C3/C4/N2), aimed at efficiently separating acid gases, such H2S and CO_2, from highly sour natural gas and other hydrocarbon-rich streams. A critical consideration in this study is minimization of hydrocarbon slippage, ensuring that valuable hydrocarbons remain within the processed gas stream, while the separated acid gases are directed to the Sulfur Recovery Unit (SRU). The membrane showed excellent and very attractive performance under harsh sour gas environment and by optimizing the membrane materials and configurations, the separation process achieves high levels of acid gas removal while maintaining minimal hydrocarbon losses. Hybrid membrane amine absorption approach also demonstrates environmental benefits by reducing CO_2 emissions significantly. Traditional amine-based acid gas removal processes require substantial energy input for the regeneration, resulting in higher CO_2 emissions. In contrast, the membrane technology operates with a considerably lower energy demand, significantly reducing the thermal duty associated with amine re-boiling. This leads to a lower overall duty for the re-boiling of the amine and thus resulting to a more promising and cost effective gas treatment process with much smaller carbon footprint.

Enhancing forward osmosis efficiency: unraveling the physicochemical interactions between Layer-by-Layer coated membranes and Pluronic® as a draw solution

Aylin Kinik (Eindhoven University of Technology)*; Marrit van der Wal (Eindhoven University of Technology); Kitty Nijmeijer (Eindhoven University of Technology); Zandrie Borneman (Eindhoven University of Technology)

Forward osmosis (FO) offers lower energy demands than traditional membrane processes like reverse osmosis; however, the main challenge is undesired draw solute transport from draw solution (DS) to feed solution (reverse draw solute flux, JS). This can be solved by tailoring the selective layer composition of the membrane for the chosen DS to minimize JS and maximize the water flux (JW).

Biodegradable amphiphilic copolymers, specifically Pluronic®, gained interest as DS due to their non-toxicity, high osmotic pressure in solution, and easy regeneration. While Layer-by-Layer (LbL) coating enables precise tuning of FO membrane properties and performance, interactions between LbL-coated membranes and Pluronic® have not yet been fully investigated.
In this research, LbL multilayers were built on hydrolysed polyacrylonitrile support (HPAN) using the strong polyelectrolytes PDADMAC (+) and PSS (-). The effect of salt concentration in the coating solution and number of bilayers (BL) was studied.

Our study revealed two types of interactions between Pluronic® and the membrane: 1) adsorption of Pluronic® unimers on and in the LbL multilayer causing pore-clogging in the multilayer and membrane support and 2) Pluronic® micelle adsorption on the LbL multilayer with the Pluronic® micelle rejection being controlled by the multilayer thickness and density. LbL coating of HPAN increased JW due to deswelling while reducing JS as a result of Pluronic® rejection by the LbL layers. Higher BL numbers increase the membrane resistance causing a decreased JW. Moreover, dense multilayers coated in 5 mM NaCl lead to multilayer removal upon Pluronic® exposure, while samples coated in 500 mM NaCl show low JW and high JS as their open multilayer structure promotes pore-clogging. The membranes coated in 50 mM NaCl show the best performance (JW = 11.6 LMH, JS = 0.2 gMH) with a very low specific reverse salt flux (JS/JW) of 0.02 g/L, proving the relevance and potential of our approach.

Mixed-gas performance, plasticization and aging behavior of thin film composite membranes based on hyper-crosslinked spirocyclic polymers

Maria Salome Tavera Villamizar (The University of Oklahoma); Lucas C. Condes (The University of Oklahoma); Matthew Webb (The University of Oklahoma); Michele Galizia (University of Oklahoma)*

The authors chose not to publish the abstract content.

Synthesis and Evaluation of Segmented Block Copolymers as Gas Separation Membranes

Elizabeth Powell (University of Tennessee)*; Rahil Parikh (University of Tennessee); Brian Long (University of Tennessee)

The development of polymers of intrinsic microporosity (PIMs) has marked a paradigm shift in the field of gas separation membranes. The rigid and kinked nature of the PIM backbone prevents polymer chain packing allowing for the formation of micropores within the polymer matrix. This allows for a molecular sieving effect that gives rise to exceptional gas transport properties for a slew of gas pairs. Notably, PIMs have been marked as a primary contributor to the revision of Robeson's upper bound in recent years. Despite the surge of research that has followed the initial development of PIM-1, rapid and extensive physical aging along with poor mechanical properties due to the brittle nature of these membranes has remained an issue. To mitigate these issues, our research aims to develop segmented block copolymers using PIM-1 and ROMP polynorbornene (PNB) to investigate the effects of adding rubbery units on gas transport properties, physical aging, and mechanical properties. To do so, we have created two series of copolymers to look at the effects of increased concentration of PIM-1 into the ROMP PNB backbone (series 1), and conversely, the effects of increased ROMP PNB into the PIM-1 backbone (series 2). Through this systematic study we have found that while the overall permeability and selectivity are expectedly decreased for both series, minimal to no physical aging is observed over a three-month period. Additionally, the tailored incorporation of ROMP PNB into the PIM-1 backbone (and vice versa) allows for tunable mechanical properties with limited effect on permeability and selectivity. Overall, we offer an additional avenue by which to tune gas transport properties while simultaneously tailoring physical aging and mechanical properties. We suspect that through simple modifications of rubbery and microporous units incorporated, these properties can be further tuned and give insight into the structure-property relationship that governs gas transport.

Tailoring Carbon Molecular Sieve (CMS) Membranes for Challenging Gas Pair Separation

DEEPANJALI ROY (WEST VIRGINIA UNIVERSITY)*; AIDAN GHEE (WEST VIRGINIA UNIVERSITY); OISHI SANYAL (WEST VIRGINIA UNIVERSITY)

In this presentation, I will describe my ongoing efforts to design membranes for the downstream separation of C2H4 from a mixture of CO, CO2 following the electrochemical conversion of CO2 to C2H4. To produce high purity C2H4, unconverted CO2 and residual CO must be removed from the product stream. CO (d = 3.76 Å) and C2H4 (d = 3.75 Å) are particularly difficult to separate due to their nearly identical sizes and similar diffusion coefficients. This project intends to make carbon molecular sieve membranes to separate CO and C2H4. CMS membranes, which are pyrolyzed end-products of rigid polymers, have demonstrated precise separations for several similar-sized gas-pairs. Their unique ability to control transport pathways allows for high diffusion and sorption selectivities, even for challenging pairs like CO/C2H4. By carefully tuning pyrolysis conditions, including temperature, soak time and ramp rates, we have tailored the intrinsic transport properties (permeation, diffusion and sorption) of these membranes to maximize their separation efficiency. The results will demonstrate how variations in pyrolysis parameters influence the micropore and ultramicropore structure and its impact on the selective transport of CO and C2H4. Additionally, I will also present results comparing polymer membranes with CMS membranes fabricated using different pyrolysis protocols.

Design principles of reactive nanofiltration membranes for water treatment

Yanghua Duan (Rice University)*; Ruoyu Wang (Rice University); Amit Shocron (Yale University); Menachem Elimelech (Rice University)

State-of-the-art water treatment technologies, such as nanofiltration and reverse osmosis, are often ineffective at removing low molecular weight neutral contaminants. Reactive nanofiltration membranes address this limitation by integrating catalytic transformation and separation processes to convert these contaminants into non-toxic products. However, their development has been hampered by limited design principles, often due to inadequate consideration of reaction kinetics and mass transport, resulting in inaccurate performance assumptions. This study employs a unified reactive transport model to assess the performance of various catalytic membranes under different operating conditions. To enhance oxidant activation on the membrane surface while minimizing excessive catalyst coating, the optimal catalyst coating density was determined by analyzing the rate-limiting step of the overall process. For membranes with catalysts coated within the active layer, the activation of oxidants could be limited by their transport to active sites, which is influenced primarily by the charge of the oxidant as well as the pore size and charge density of the membrane. While selective solute rejection enhances contaminant transformation within the membrane, it has minimal impact on surface reactions. In addition to membrane design, treatment performance can be improved by selecting oxidant-catalyst pairs that generate multiple reactive species. Overall, the proper design of reactive membranes ensures effective contaminant removal and salt rejection under diverse operational and influent conditions. This study provides a theoretical framework for designing and optimizing reactive nanofiltration membranes for advanced water treatment applications.

Composite graphene oxide hollow fiber membranes for hydrogen separation

Guilherme Ascendino (Universidade Federal de Uberlândia); Carla Hori (Universidade Federal de Uberlândia); Miria Reis (Universidade Federal de Uberlândia)*

Graphene oxide membranes have been largely studied for gas separations. However, the assembly of graphene oxide sheets on a porous substrate is a challenge to obtain highly selective membranes. Here we evaluate the application of graphene oxide membranes that were deposited on alumina hollow fiber for hydrogen separation. The ceramic hollow fibers were produced by the phase inversion process followed by a single sintering step that ensured an asymmetric pore size distribution to the produced substrate. After the deposition, the graphene oxide membranes were reduced at a mild temperature of 160 °C to reduce the interlayer space between the graphene oxide sheets and improve the membrane selectivity. According to XRD analyses, the interlayer space decreased from 8.71 to 7.38 Å after the thermal annealing process at 160 °C. The composite graphene oxide membrane presented an H2 permeance of 2032.47 ± 101.69 GPU and an H2/N2 selectivity of 12.16 ± 0.30. Thus, the produced composite membranes presented suitable characteristics to be applied for gas separations.

Clarification of natural extracts by ceramic hollow fiber membranes

Kristopher Dorneles (Universidade Federal de Uberlândia); Karolyna Miguel (Universidade Federal de Uberlândia); Vicelma Cardoso (Universidade Federal de Uberlândia); Miria Reis (Universidade Federal de Uberlândia)*

The application of ceramic membranes for clarifying natural extracts is recommended to preserve the thermally sensitive bioactive compounds. Here we evaluate the application of ceramic hollow fibers to concentrate the bioactive compounds from clove basil leaf and annatto seed extracts. These neglected and underutilized species (NUS) present high concentration of bioactive compounds. Figure 1(b) shows the visual appearance of the feed, concentrate and permeate samples. Filtrations of the annatto seed extract were carried out through a kaolin hollow fiber membrane at a transmembrane pressure of 1.5 bar. The steady state flux for a concentration factor of 1.5 was almost 20 L/(h m2). Kaolin hollow fiber membranes were able to concentrate 26% of the bixin molecules and 18% the phenolic compounds from the annatto seed extract. The clove basil leaf extract was concentrated by a spinel hollow fiber membrane and the steady state flux was also of approximately 20 L/(h m2). After filtration of the clove basil leaf extract, the permeate sample retained almost 60% of its phenolic content and more than 65% of its antioxidant potential. Thus, the ceramic hollow fiber membranes were suitable for clarifying aqueous plant extracts, maintaining the antioxidant capacity and health-promoting compounds.

The application of tangential flow filtration for the continuous processing of precipitated monoclonal antibodies: process optimization in screened membrane cassettes

Zach Badinger (Penn State University); Ali Behboudi (Penn State University); Andrew Zydney (Penn State University)*

Recent advances in cell culture have led to significant increases in monoclonal antibody (mAb) titer, opening a new window of opportunity for developing fully continuous downstream purification processes for mAb manufacture. One approach is to selectively precipitate the mAb from harvested cell culture fluid using ZnCl2 and PEG, with the protein precipitate separated from impurities such as host cell proteins using single-pass tangential flow filtration (SPTFF) with microfiltration membranes. Previous studies have proved that particle morphology has a significant impact on the filterability of the precipitated mAb. In this regard, employing certain salts such as sodium malonate and CaCl2 as densifying agents leads to significantly larger values of the critical flux. The objective of this work was to examine the potential of obtaining further improvements in SPTFF by using screened membrane cassettes instead of the hollow fiber membranes that have been used in almost all previous studies. The combination of sodium malonate, to increase the density of the precipitate, and the enhanced mass transfer characteristics of the screened cassette resulted in almost a doubling of the critical flux, which was evaluated from flux-stepping experiments. This increase in critical flux provided a large increase in the single pass conversion (ratio of permeate to feed flow rates), enabling much higher concentration factors and enhanced impurity removal. Further improvements in performance were obtained using two membrane cassettes in series, allowing operation at higher feed flow rates. Data obtained during a continuous 24-hour filtration showed minimal fouling at a filtrate flux of 144 LMH with over 80% conversion. These results demonstrate the potential of using screened membrane cassettes for the continuous / intensified processing of precipitated monoclonal antibodies and other high-value protein products.

Structure-Property-Performance Relationships of Upcycling Waste PMMA in PVDF Membrane Design

Joseph Danner (The University of Alabama)*; Liam Papier (The University of Alabama); Mia Eisenberg (The University of Alabama); Pravin Shinde (The University of Alabama); Steven Weinman (The University of Alabama)

With an annual production of ~4 million metric tons, polymethyl methacrylate (PMMA) is a widely used transparent thermoplastic for many applications. Estimates indicate that only 10% of waste PMMA is recycled, while the majority is disposed of by conventional means. A promising strategy to divert this waste from landfills is to utilize solution blending to produce value-added materials for membrane fabrication by leveraging the well-studied miscibility between PMMA and polyvinylidene fluoride (PVDF). This study probes the structure-property-performance relationship of PVDF/PMMA blends to understand how different additive and molecular weight compositions in waste PMMA sources affect PVDF membrane properties. Laboratory cuvettes and fiber-optic cables were used as sources of PMMA waste. 16 wt% PVDF/PMMA blends were prepared with two greener solvents, dimethyl sulfoxide and triethyl phosphate, to produce flat-sheet membranes using nonsolvent-induced phase separation (NIPS). Waste plastics were evaluated with spectroscopic and chromatographic techniques to characterize polymer properties and to determine if processing additives or co-polymers were present. Membrane formation kinetics were quantified by measuring the nonsolvent bath total organic carbon concentration as a function of time during fabrication. Scanning electron microscopy (SEM) was used to evaluate the cross-sectional and top surface morphologies of fabricated membranes. Pure water permeance and solute rejection were quantified with dead-end filtration. We found that increasing the PMMA loading from 0 to 4 wt% proportionally decreased the permeance and surface porosity from 170 ± 18 to 75 ± 8 LMH/bar and 34 ± 5 to 6 ± 1%, respectively. However, the separation performance significantly improved for the entire range of macromolecular solutes (20 – 200 kDa). These results indicate that incorporating waste PMMA into PVDF membranes produces distinctly different membranes with tailorable separation properties.

Membrane-Assisted Entrainer Strategy for Efficient and Sustainable Purification Processes

Seung Hwan Kim (Seung Hwan Kim)*

Efficient separation and purification of chemically similar compounds remain significant challenges in various industrial applications, particularly in biorefineries and ion recovery processes. The Entrainer Strategy, inspired by azeotropic distillation, introduces selective binding agents to enhance separation performance by altering target molecules' effective size or charge. This present demonstrates the versatility and scalability of membrane-assisted entrainer strategies for the sustainable purification of ionic compounds (e.g., Lithium, Nickel, Organic acids, etc). Using innovative entrainers such as protonated polyethyleneimine (PEI)[1], ionophore, and poly(4-styrene sulfonic acid) (PSS)[2], this strategy achieves selective binding through electrostatic interactions, hydrogen bonding, or coordination chemistry. Key examples include the purification of organic acid, recovery of lithium ions from sodium-rich environments, and separation of nickel ions from brine solutions.

By integrating entrainer-assisted separation into membrane cascade systems, significant improvements in purity (up to 99.9%) and yield (exceeding 80%) were achieved, overcoming traditional limitations of nanofiltration and ultrafiltration. Techno-economic analyses revealed that the membrane-assisted entrainer strategy reduces energy consumption by up to 50% compared to conventional ion exchange or distillation-based processes, with a return on investment (ROI) of less than one year in large-scale operations. Furthermore, entrainer regeneration and recyclability were validated across multiple cycles without performance degradation, emphasizing the sustainability of this approach.

This work represents a paradigm shift toward scalable, energy-efficient, and sustainable membrane operations, providing innovative solutions for resource recovery and advanced separation technologies.

Reference

[1] Chemical Engineering Journal 493, 152247.
[2] Journal of Membrane Science 715, 123442.

INFLUENCE OF SURFACE PATTERNING COMBINED WITH MOF GRAFTING ON OIL FOULING AND SELENIUM REMOVAL DURING NANOFILTRATION

Lauren Ward (The University of Alabama)*; Sweta Modak (The University of Alabama); Catherine Martin (The University of Alabama); Milad Esfahani (The University of Alabama); Steven Weinman (The University of Alabama)

Surface patterning is an interest to the membrane community after the success seen of the Sharklet® pattern, which mimics the form and function of shark skin, in the biomedical industry. The surface patterns disrupt fluid flow at the membrane surface that encourages localized mixing that can prevent fouling and creates an easier to clean membrane. Fouling leads to a decrease in rejection and water transport through the membrane while also leading to a decrease in the membrane lifetime. In previous work, it was found that patterns that are larger than or smaller than the foulant size can lead to a decrease in fouling rate. In this project, we investigate how patterned membranes with metal organic frameworks (MOFs) attached to the surface effect fouling in nanofiltration with ~10-micron sized foulants. Pristine and 50-micron patterned membranes will be tested as is along with MOF-808 attached to the surface in a cross-flow system in order to determine the effectiveness of adding the MOF to the surface to reduce fouling. The 4 different membranes will be tested with 2 oil solutions in order to determine this. The hypothesis of this project is that by attaching the MOFs to the surface of the membrane, there will be more localized turbulence and thereby having an increase in flux recovery.

Membrane Preparation by Reticulate Rational Design

Ben Bikson (Avanpore)*; Ninad Khadse (Avanpore, LLC); Sagiv Weintraub (Chromodynamix, LLC)

The next generation of high-tech products demands increased raw materials and product purity while deploying green manufacturing processes. Industrial chemical and biological process streams are mixtures of molecules that require extensive separation and purification on a molecular level to isolate the product. Membrane technology is commonly used in industries as diverse as water treatment, desalination, pharmaceuticals, and electronics. Deployment of membranes to treat industrial streams has been limited due to the chemical solvent resistance of commercially available membranes and molecular-level separation efficiency. The process streams are frequently mixtures of corrosive fluids and/or organic solvents. Membrane processes can be limited to the treatment of water-based streams by chemical solvent resistance of membrane materials.

Composite membranes were prepared by Reticulate Rational Synthesis (RRS). The method was used to prepare chemically/solvent-resistant composite membranes using poly (ether ether ketone), PEEK, hollow fibers as a substrate. Hollow fibers formed from PEEK polymer exhibit almost universal solvent/chemical resistance and outstanding thermo-mechanical properties. These properties make PEEK an ideal material for the preparation of membranes with controlled separation characteristics by hierarchical surface modification. Asymmetric PEEK hollow fibers were used as a substrate for the preparation of composite membranes with Janus structure and separation characteristics in the UF, and NF range. The composite membranes are formed without affecting the preformed substrate pore structure. RRS methodology provides for the formation of a selective separation layer via direct chemical modification, grafting, and layer-by-layer synthesis with tailored pore size and functionality. Composite hollow fiber membranes are formed with the separation layer covalently attached to the substrate providing a robust chemical/solvent-resistant structure.

A Process-Structure-Property Modeling and Optimization Framework for Membrane-based Selective Nutrient Recovery

Md Masud (West Virginia University)*; Km Piash (West Virginia University); Oishi Sanyal (West Virginia University); Yuhe Tian (West Virginia University)

The discharge of ammonium (NH_4^+) and potassium (K^+) ions from wastewater is a significant environmental challenge, driving nutrient pollution that leads to harmful algal blooms and eutrophication, harming ecosystems and incurring high costs. Membrane-based processes offer significant potential for recovering nutrients from wastewater-derived anaerobic digestates for agricultural fertilizers.

This study introduces a novel computational framework for process-structure-property modeling and optimization of membrane-based nutrient recovery processes. The framework integrates three components: membrane manufacturing, structure-property modeling, and systems-level optimization. A quantitative structure-property relationship links the crosslinking density of glutaraldehyde (GA) to the rejection performance of NH_4^+ and K^+ ions in surface-modified nanofiltration membranes. Data-driven modeling is used to address the lack of mechanistic models and understand the influence of GA concentration on membrane separation properties.

Building on this, we design and optimize a continuous membrane-based intra-nutrient separation process. Key considerations include mass balances, selective separation driving forces, permeance as a function of GA concentration, membrane cost correlations, and nutrient flow rate targets. A superstructure network representation of membrane modules (parallel, sequential, and hybrid configurations) enables systematic exploration of design alternatives. A mixed-integer nonlinear optimization model minimizes membrane surface area costs while meeting operational and performance constraints. The objective function also addresses membrane manufacturing parameters and trade-offs between numbering-up and scaling-up strategies. This work advances sustainable wastewater management and circular economy goals by transforming nutrient recovery into an environmentally and economically viable process.

Investigating the Impact of Temperature and Concentration on Membrane-Assisted Crystallization of Lithium Fluoride Solution

Giuseppe Prenesti (University of Calabria)*; Mustafa Ali (University of Calabria); Alessio Caravella (University of Calabria); Francesca Macedonio (ITM-CNR); Alfredo Cassano (ITM-CNR); Elena Tocci (ITM-CNR)

Lithium fluoride (LiF) is a versatile compound widely used in optics, ceramics, nuclear technology, and lithium-ion batteries due to its excellent electrochemical stability and ionic conductivity. In batteries, LiF forms stable solid electrolyte interphases (SEI), improving performance, efficiency, and longevity. However, the extensive use of batteries raises sustainability concerns due to waste and pollution, driving attention to recycling as a solution for environmental and economic challenges (Ramirez et al., 2024; Battaglia et al., 2022). Recovering critical metals like lithium and cobalt has become essential, with reactive crystallization techniques emerging as key solutions. Membrane-assisted crystallization (MAC) has shown great promise in lithium recovery, utilizing hydrophobic membranes like polypropylene (PP) to selectively transport water vapor while retaining lithium salts, enabling controlled supersaturation and crystallization. This process enhances recovery efficiency and aligns with circular economy principles (Alessandro et al., 2024). Our study uses molecular dynamics (MD) simulations to explore LiF crystallization behavior in MAC systems (Prenesti et al., 2025). Simulations conducted with GROMACS 5.1.4 (Abraham et al., 2015) and VMD 1.9.3 (Humphrey et al., 1996) modeled water using the SPC/E model and ions with the OPLS force field. Systems at varying concentrations were equilibrated under NPT ensembles, followed by 300 ns production runs at different temperatures and 1 atm. Results highlight how hydrophobic membranes influence nucleation and crystal growth under different supersaturation conditions, demonstrating their role in stabilizing nuclei and shaping crystal morphology. These insights pave the way for efficient and sustainable lithium recovery technologies.

Ethanol Vapour-Induced Morphological Influences on the Water Barrier Performance of PLA Films: A Multi-component Study.

Sean McIntyre (surface measurement systems)*; Dian Xu (Michigan State University); Samantha Adjetey (Imperial College London); Rafael Auras (Michigan State University); Daryl Williams (Surface measurement systems); Paul Iacomi (Surface measurement systems)

Poly-lactic acid (PLA) is a widely used biopolymer in the healthcare industry due to its inherent biocompatibility and good mechanical properties. In healthcare applications, PLA can come into contact with water/ethanol mixtures during sterilization or be exposed to humid environments which favor PLA degradation. In this work, a multi-component permeation study to investigate the permeation of water/ethanol vapor mixtures through the PLA films at varying concentrations and temperatures, in order to reproduce real-world conditions. Vapors were introduced to the top of the membrane and the permeate measured underneath. Further studies were undertaken using dynamic vapor sorption – Raman techniques to study water and ethanol vapor uptake, and the associated changes in crystallinity.

Preliminary results are consistent with those reported by Auras et al. where the water vapor permeability coefficient (WVPC) decreases through PLA with temperature, which was attributed to the relaxation of the rigid amorphous fraction (RAF) present at the boundaries of crystalline and amorphous domains. When the WVPC is analyzed in the presence of 25% ethanol, we observe an overall increase in the WVPC (~40-60%) and a shift in the temperature at which the RAF contributes to a decrease in permeability (from 35 °C to 45 °C). This shift in the temperature effects is also seen in Iniguez-Franco et al. where the addition of an ethanol fraction of 25% resulted in a reduction of the glass transition temperature from >50 °C to >43 °C. The reduction of the Tg was attributed to solvent-induced crystallinity over long periods, whereas in this study it is likely to be a swelling effect over the smaller time scales. Investigations into the partition coefficients revealed a decrease (~25%) in the water partition coefficient with increasing ethanol concentrations, suggesting that ethanol decreases the solubility of water in PLA which results in a lower permeability.

Metallic Membrane Reactors for Blue Hydrogen Production

Hani Hawa (Precision Combustion, Inc.)*; Subir Roychoudhury (Precision Combustion, Inc.)

Inorganic membrane technology is being investigated to perform many of the separations needed for H2 generation from fossil fuels reforming or gasification while simultaneously concentrating and capturing CO2. This talk will focus on high temperature, metallic-based, hydrogen permeable membrane materials, such as palladium (Pd) or Pd-alloys and their integration in catalytic membrane reactors.

Pd and its alloys are the most widely studied due to their high H2 permeability, their chemical compatibility with many hydrocarbons containing gas streams, and their theoretically infinite H2 selectivity. Our approach has been to fabricate composite membranes consisting of thin, micron-thickness, Pd-alloy films on tubular, porous ceramic substrates using proprietary deposition techniques. Trade-offs associated with several Pd-alloy composition and permeation data with pure and mixed feed gases at high pressures and temperatures will be presented. Examples of current and developmental process applications of Pd alloy membranes will also be discussed where H2 production can be coupled with CO2 capture. These applications include H2 recovery from gasification-derived synthesis gas containing H2S and a membrane-assisted reactor to produce H2 via steam reforming of methane while simultaneously concentrating the CO2 product. Levelized Cost of H2 (LCOH) analysis data will also be presented and benchmarked against DOE's metrics.

Pillared Laminar Phyllosilicate Membranes with Tunable Ion Sieving

Yining Liu (University of Chicago)*; Yuqin Wang (University of Chicago); Bratin Sengupta (Argonne National Laboratory); Alex Martinson (Argonne National Laboratory); Jeffrey Elam (Argonne National Laboratory); Seth Darling (Argonne National Laboratory)

Membranes incorporating two-dimensional (2D) materials have shown great potential for water purification, resource recovery [1], and energy storage and conversion applications to overcome the challenges of traditional polymeric membranes. A variety of 2D materials such as graphene oxide, transitional metal dichalcogenides, and MXenes have been investigated for membrane applications. Phyllosilicates are naturally occurring 2D materials that are earth-abundant and low-cost, which have been introduced as abundant and cost-effective source materials for such membranes. However, their water instability and low inherent ion transport selectivity hinder practical applications. In previous work, we have demonstrated diamine crosslinked phyllosilicate membranes through electrostatic interaction for enhanced stability [2]. In this work, we report a new class of laminar membranes with excellent stability and ion transport selectivity by incorporating inorganic alumina pillars into vermiculite interlayers. Covalent crosslinking with pillars significantly enhances stability by strengthening interlamellar interaction. Doping of cations during the pillaring process reverses the membrane's surface charge, substantially boosting monovalent/multivalent cations through electrostatic interactions. Through further introduction of cations after the pillaring process, separation between monovalent cations is also enhanced by size exclusion. This work realizes both monovalent/multivalent and monovalent/monovalent selective ion sieving with the same membrane platform.

[1] O. Kazi et al., Adv. Mater. 35 (2023) 2300913
[2] Y. Liu et al., ACS Appl. Mater. Interfaces 15 (2023) 57144

Specific Ion Effects on Ion Transport in Charged Polymer Membranes

David Kitto (University of Michigan)*; Gregory Reimonn (University of Michigan); Riley Vickers (University of Michigan); Carolina Espinoza (University of Michigan); Bryan Goldsmith (University of Michigan); Jovan Kamcev (University of Michigan)

Selective separation of like-charged ions remains a critical challenge in applications such as water treatment and resource recovery. Electrochemical membrane-based separations offer a promising pathway to address this challenge, but a limited fundamental understanding of ion transport in charged polymer membranes hampers the development of highly selective materials. Here, we discuss the role of specific ion effects (SIEs) on ion transport through water-swollen charged polymer membranes (ion-exchange membranes or IEMs). Our investigation integrated experimental measurements of ion mobility for 15 different ions with all-atom molecular dynamics simulations to gain deeper insight into the molecular interactions underpinning SIEs. To unite the experimentation and simulation methods of investigation, we performed in situ measurements of ion/ion and ion/water interactions via Raman spectroscopy and differential scanning calorimetry, respectively. Altogether, these data demonstrate that often-invoked membrane phenomena such as contact ion pairing and ion dehydration do not explain the different energetics of monovalent ion transport across the membranes. Instead, solvent-mediated ion interactions drive pronounced SIEs within the studied membranes. The ion softness, i.e., the malleability of the ion hydration shells, from Hard/Soft Acid/Base (HSAB) theory emerged as a key predictor of transport energetics and simulated interactions. Impressively, HSAB theory not only describes the quantitative differences in ion transport within this dataset, but also predicts the deduced mechanism of solvent-mediated ion interactions. These findings begin to construct a mechanistic framework for designing membranes with tailored ion selectivity, which could enable the efficient separation of chemically similar ions.

Theory for Dynamic Ion Transport in Ion-Shuttling-Electrodes for Electrochemical Ion Pumping

Weifan Liu (Vanderbilt University)*; Jouke Dykstra (Wageningen University & Research); Maarten Biesheuvel (Wetsus); Longqian Xu (Vanderbilt University); Shihong Lin (Vanderbilt University)

Electrochemical ion pumping (EIP) is an innovative electric field-driven separation process that combines the advantages of conventional capacitive deionization and electrodialysis for effective ion separations. The composite electrode used in EIP combines an ionic conductor in the form of ion-exchange material, with a finely dispersed matrix of electrically conductive porous carbon particles for ion storage. This study presents a comprehensive mathematical model for dynamic ion transport in an EIP process with ion-exchange polymer-filled porous carbon electrodes in the context of desalination. The model integrates the classical Nernst-Planck equation for ion transport within the ion-exchange polymer and an extended Donnan model for ion storage in the porous carbon. By analyzing spatiotemporal distributions of ion concentrations and potentials in the charging and discharge steps, the model reveals interesting dynamic behavior of ion transport in EIP that is very different from that in conventional capacitive deionization or electrodialysis. Validation of the ion transport model is performed using EIP experiments with constant current and a cell with a single cation shuttling electrode. The model successfully captures the dynamic changes of electrode potential and salt concentrations in the flow channel under different experimental conditions. The model lays the theoretical foundation for EIP, which is critical to future work on modeling large-scale EIP stacks and EIP for selective ion separation, as well as performing model-based optimization of system design and operation.

LOCALIZED SURFACE PLASMON RESONANCE (LSPR) FOR PREDICTING FOULING AND SCALING IN RO MEMBRANES DURING SEAWATER DESALINATION

Iniya Prasanthi (Ben-Gurion University of the Negev); Mahaveer Halakarni (Ben-Gurion University of the Negev); Moshe Herzberg (Ben Gurion U)*

Predicting fouling and scaling in reverse osmosis (RO) systems is essential for maintaining efficiency and reducing operational challenges. Fouling, caused by organic, inorganic, and biological matter, and scaling, particularly from silica, disrupts system performance by reducing flux, increasing energy demands, and shortening membrane lifespan. Silica, with its low solubility and hard deposits, is especially problematic.1 Current methods like the Silt Density Index (SDI) and turbidity analysis offer limited accuracy and lack real-time monitoring capabilities.2 Localized surface plasmon resonance (LSPR) sensors provide a promising alternative, offering precise, real-time detection of fouling agents with enhanced sensitivity. By enabling better pre-treatment optimization and early intervention, LSPR has the potential to improve membrane longevity, reduce maintenance costs, and advance sustainable water treatment practices.

The LSPR sensor was modified to mimic commercial RO membranes by introducing a cross-linked aromatic polyamide layer. Adsorbed mass surface concentrations of foulants (e.g., Bovine Serum Albumin, Humic Acid, Sodium Silicate) and synthetic seawater dual foulants were quantified by monitoring shifts in the light-extinction maxima ($\Delta\lambda$), corresponding to refractive index changes caused by foulant accumulation (Fig 1). LSPR results were compared with SDI to assess accuracy.

The LSPR sensor's high sensitivity enables real-time fouling detection, providing a powerful tool to enhance RO system management and longevity.

High Recovery RO-NF Desalination of Groundwater of High mineral Scaling Potential

Yakubu Jarma (University of California, Los Angeles)*; Beau Akhavain (University of California, Los Angeles); Yoram Cohen (Professor)

High recovery inland brackish water desalination is critical to attain high permeate productivity and reduce concentrate management challenges. However, high recovery desalting is difficult for source water of high mineral scaling potential, even with antiscalant dosing. At high recovery, membrane scaling due to mineral salt crystallization limits permeate productivity, increases operational costs, and shortens membrane lifespan. To address these challenges for inland water of high calcium carbonate scaling potential, a novel integrated treatment train was evaluated consisting of: (i) a flexible system combining RO and NF elements (FLERONF) for primary (PRONF) high-recovery desalting that includes concentrate recycling, (ii) PRONF concentrate desupersaturation via accelerated seeded crystallization, and (iii) FLERONF secondary desalting (SRONF) for recovery enhancement. The FLERONF system (~1 m^3/h permeate capacity) included two 2.5-inch RO elements and one 2.5-inch NF element in series. Brackish groundwater (1,200 mg/L TDS) was desalted in the PRONF stage at 70% recovery, producing a concentrate of ~3,800 mg/L TDS. Concentrate desupersaturation, with respect to calcium carbonate, was conducted at alkaline pH (8–10.5) along with calcite seeds to accelerate precipitation. Batch crystallizer tests determined the optimal lime dosing for antiscalant deactivation to suppress interference with precipitation. Desupersaturation reduced the calcite supersaturation index of PRONF concentrate from 168 to 10, enabling further desalting with antiscalant dosing. The microfiltered, desupersaturated PRONF concentrate had a turbidity of <<1 NTU, facilitating SRONF desalting which enhanced the overall recovery to >90%. The above approach integrated desalination train reduce the levelized water costs by minimizing membrane scaling and concentrate management costs, demonstrating an effective strategy for high-recovery inland desalting.

Enhancing the performance of vacuum membrane distillation (VMD) membrane through solvent crystallization with a bi-solvent system during phase inversion

Nguyen Lam (Nanyang Technological University)*; Paulo Jorge Da Silva Bartolo (Nanyang Technological University); Chong Tzyy Haur (Nanyang Technological University)

Membrane distillation (MD) technology is a promising membrane separation process for seawater desalination and wastewater treatment due to its compatibility to renewable energy, thermal efficiency, and ability to treat highly saline feed. For MD application, it is essential to have a membrane that is resistant to pore wetting and possesses high flux. Previous research work has focused on modifying the membrane with fluorosilanes and nanoparticles to increase its hydrophobicity. In this study, to achieve high flux and wetting resistant membrane suitable for vacuum membrane distillation (VMD), a novel strategy of using the bi-solvent system of dimethyl sulfoxide (DMSO) and dimethyl sulfone (DMSO2) was adopted to trigger solvent crystallization during phase inversion in icy water bath and create connected pore structures. Since DMSO and DMSO2 has different freezing point (at 190C and 109oC respectively), the ratio of DMSO2 to DMSO determines the final freezing point of the bi-solvent system. Optimizing the solvent ratio helps to facilitate the solvent crystallization process, minimize the dead-end pores and control the pore size of the membrane. Using this method, a poly(vinylidene fluoride-co-hexafluoropropylene) (PVDF-co-HFP) top layer was fabricated onto a polyethylene terephthalate (PET) non-woven support layer. The fabricated membranes have a pore size range of 30 to 50 nm, an active layer thickness of 205±10 um and liquid entry pressures (LEP) range of 8.6 to 12.8 bar. The optimum membrane attained from the bi-solvent system had a stable flux of 12.0±0.4 L/m2h (an over 10-fold increase in flux as compared to the membrane obtained from the single DMSO solvent system) and a salt rejection of over 99.99% when tested in a VMD set-up with 35g/L of NaCl for 7 hours.

High-Pressure Reverse Osmosis for Continuous Downstream Concentration of Bio-Based Carboxylic Acids

Yian Chen (KGI)*; Hakan Olcay (NREL); Eric Tan (NREL); Gregg Beckham (NREL)

The production of bio-based chemicals through fermentation and enzymatic processes often presents significant challenges in product separation, which tend to be the most costly and energy-intensive steps in the entire process. This is especially true when dealing with low concentrations of target bioproducts. In this study, we investigated the use of high-pressure reverse osmosis (HPRO) as a method for concentrating an exemplary fermentation product, butyric acid—an important precursor for sustainable aviation fuel—prior to downstream extraction. Through a combination of modeling and experimental analysis, we identified the key factors limiting the maximum achievable concentration factor (CF) of butyric acid to 4.0 with an HPRO membrane, representing a notable improvement over the 2.6-3.2 CF range typical of conventional reverse osmosis (RO) membranes.

Following concentration, the aqueous stream underwent liquid-liquid extraction using an organic solvent, followed by distillation for butyric acid purification and solvent recovery. The incorporation of HPRO for product concentration into an in situ product recovery (ISPR) process resulted in a more than five-fold increase in the final butyric acid concentration in the organic phase. Additionally, this approach led to a substantial 76% reduction in organic solvent consumption. These advancements contributed to a significant reduction in both production costs and environmental impact, with a 53% decrease in the overall cost of butyric acid production and a 46% reduction in greenhouse gas (GHG) emissions. These improvements far outpace the performance of conventional RO membranes in product concentration.

Overall, the integration of HPRO membranes for product concentration offers a more economically viable and environmentally sustainable approach to the recovery of bioproducts from dilute aqueous streams, providing clear benefits in both process efficiency and sustainability.

Turning Aquaculture Waste into Fertilizer and Clean Water: Optimizing Membrane Distillation for N and P Recovery

Erica Bertozzi (Politecnico di Torino)*; Gabriele Copetti (Politecnico di Torino); Lorenzo Craveri (Politecnico di Torino); Ali Naeimi Tabasian (Politecnico di Torino); Marco Malaguti (Politecnico di Torino); Alberto Tiraferri (Politecnico di Torino)

Aquaculture wastewaters (AQWs) are rich in dissolved nutrients, mainly nitrogen (N) and phosphorus (P), offering potential for resource recovery. Membrane distillation (MD) has emerged as a promising technology to recover N as liquid ammonium fertilizer and enable P precipitation as struvite, a solid fertilizer, when a magnesium source is available. However, balancing both nutrient recoveries requires tuning of the ammonia (NH3) flux to prevent excessive N depletion and to maintain optimal conditions for struvite formation. In this study, a multi-step approach is proposed using direct contact MD (DCMD) to enhance N recovery while ensuring the conditions for struvite precipitation.

A cross-flow lab-scale system with a PVDF commercial membrane (140 cm2 active area) was used to assess DCMD performance under various operating conditions. Synthetic feed solutions and real AQWs (saline and freshwater) were tested. NH3-to-water vapor selectivity was adjusted based on the feed composition to achieve NH3 recovery rates of 30-70% in the distillate stream. Through a 2-D transient model, pH and temperature gradients were identified as the primary factors influencing NH3 flux. With a feed pH of 7-8.5 and a temperature of 50°C, P was concentrated up to 5 times in the retentate stream while ensuring a sufficient N content for struvite precipitation, which was obtained by raising the concentrate pH to 9. Real AQWs required pretreatments (e.g., calcium precipitation) to reduce matrix interferences. The diluted NH3-solution from partial NH_3 passage in the first MD stage was further concentrated up to 6 times in a second MD stage at pH < 6.5, adding C6H8O7 as an acidic collector to prevent NH_3 loss to the distillate. As a result, ammonium citrate fertilizer was produced while reclaiming up to 60% of the waste volume as clean water. Overall, this study demonstrates a robust MD-based strategy that allows the simultaneous production of high-value fertilizers and clean water from AQWs.

Deciphering the Impact of Transmembrane Pressure on Reverse Osmosis Composite Membrane Compaction and Performance

Jishan Wu (Rice University)*; Menachem Elimelech (Rice University); Ying Li (University of Wisconsin-Madison); Eric Hoek (UCLA)

This study examines the compaction behavior of thin-film composite reverse osmosis (TFC RO) membranes, emphasizing the roles of transmembrane pressure (TMP) and water flux. Using a crossflow system with a constant hydraulic pressure of 60 bar and varying feed solution concentrations to control TMP, results demonstrate that higher TMP and water flux significantly intensify compaction. Both commercial and hand-cast membranes experienced substantial compaction at high TMP (25–30% at 50 bar TMP) compared to less than 10% at 10 bar TMP. Scanning electron microscopy (SEM) highlighted a strong correlation between TMP and compaction in the polysulfone support layer, while molecular dynamics (MD) simulations revealed decreased porosity and reduced polyamide (PA) active layer thickness under high TMP. Real-time animations further illustrated varying degrees of PA layer compaction with TMP changes. Wet-testing and MD simulations confirmed that high TMP exacerbates compaction in both active and support layers, resulting in reduced water permeance, increased water flux, enhanced salt rejection, and elevated salt permeability. Testing order (from high TMP to low TMP or vice versa) showed no significant effect on performance due to the relaxation of the PA active layer after pressure release, while the irreversible compaction of the support layer had a limited impact. Flux dependency was evident as higher water flux induced greater compaction, with hand-cast membranes showing lower water flux (48 LMH) than SWRO (57 LMH) and BWRO (63 LMH) at 50 bar TMP. These findings provide critical insights into membrane compaction mechanisms, guiding the optimization of reverse osmosis membrane performance and the development of next-generation technologies.

Technology for Cyclic Siloxane Removal: Utilizing Hollow Fibers for Sustainable Silicone Oil Purification

Jinyoung Kim (Georgia Tech)*; Jay Rose (Dow Chemical Company); Dimi Katsoulis (Dow Chemical Company); Reza Panah (Dow Chemical Company); Ryan Lively (Georgia Tech); William Koros (Georgia Tech)

In the production of polydimethylsiloxane polymers, residual cyclic siloxanes such as D4, D5, and D6 often persist in the product stream due to the synthesis methods. Although previously tolerated in the stream, recent environmental regulations in some geographies now necessitate their removal for some market applications. Addressing the energy-intensive nature of traditional impurity removal methods such as stripping and volatilization, we describe a novel and energy-efficient membrane separation technique. This method utilizes a hollow fiber, distinguished by its high selectivity and productivity. This technique strategically employs a coating layer on the fiber that allows passage of low molecular weight cyclic siloxanes while effectively excluding higher molecular weight silicone oil polymers. By processing crude silicone fluids laden with siloxane impurities, this method demonstrates a substantially higher efficiency compared to traditional separation approaches. This approach underscores the significant potential of this method for advancing purification technologies within various industrial contexts. The technique also challenges conventional separation paradigms by offering a sustainable and efficient alternative.

Adhesion behavior in micro/ultra-filtration membranes bonded under industry-relevant conditions

Rajan Jain (Membrane Science, Engineering, and Technology (MAST) Center, Department of Mechanical Engineering, University of Colorado Boulder)*; Jaylene Martinez (University of Colorado Boulder); Riccardo Lovison (University of Colorado Boulder); Christina Carbrello (MilliporeSigma); Kathy Youngbear (MilliporeSigma); Sean Foley (MilliporeSigma); Rong Long (University of Colorado Boulder); Yifu Ding (University of Colorado Boulder)

Understanding adhesion behavior of components during membrane device manufacturing is important to achieve reliability and durability. Among different adhesion methods, thermomechanical bonding is commonly used to bond porous membrane with a thermoplastic substrate. Little is known about the adhesion mechanism between the membrane and substrate under thermomechanical loading. In this systematic study, polypropylene (PP) substrates and polyethersulfone (PES) membranes are thermomechanically bonded in industrially-relevant time scales. The effects of bonding pressure (1 MPa and 7 MPa), and membrane characteristics including pore size (20 nm and 200 nm) and surface chemistry (acrylamide- and acrylate-modified) on adhesion strength between PP and PES were examined by T-peel tests and image analysis. The dominant debonding mechanism between bonded PP and PES was shown to be interfacial failure. The interfacial fracture toughness of the 200 nm PES/PP samples (100-200 N/m) was significantly greater than that of the 20 nm PES/PP samples (10-100 N/m). Surprisingly, bonding pressure and surface chemistry imparted little-to-no impact in the adhesion strength of the 200 nm PES/PP samples but exhibited a clear reduction in that of the 20 nm PES/PP samples. The interfacial failure of the infiltrated PP fibers was hypothesized to be the main contributor to the measured interfacial fracture toughness. A theoretical estimation of the energy required to break the PP fibers was within the range of the experimental values of the interfacial fracture toughness.

Utilization of Membrane Technologies for Sustainable Peptide Synthesis and Purification

Endras Fadhilah (University of Kentucky)*; Emily Ingram (University of Kentucky); Malgorzata Chwatko (University of Kentucky)

Peptides therapeutics have seen a significant increase in demand, with many individuals benefiting from therapies such as GLP-1 agonists like Mounjaro®. However, many current technologies for peptide synthesis have run into difficulties in keeping up with the growing demand. Scale up issues are due in part to limitations with peptide synthesis occurring on a column support. A proposed switch to using liquid phase synthesis methods on soluble polymeric molecules such as poly(ethylene glycol), has led to a burst of activity on the development of sustainable and scalable approaches to peptide synthesis in solution. The goal of our work is to evaluate the use of organic solvent-based ultrafiltration for the purification steps in peptide synthesis, specifically using propylene carbonate as a more sustainable solvent.

Normal flow pressure-driven filtration experiments were conducted to assess permeability and compatibility of the organic solvents with the membrane. Several membranes with different molecular weight cutoffs (MWCO) were evaluated including 3, 10, and 30 kDa regenerated cellulose (RC) membranes and 500 and 900 Da silicone membranes. The results showed that at 40 psi, 10 kDa RC membrane was able to retain 91.3% of multi-arm poly(ethylene glycol) anchor on which peptides would be synthesized on. Additional studies noted that chemicals necessary for peptide synthesis did not alter membrane performance ensuring membrane performance remained high during peptide synthesis. Lastly, a tripeptide (Leu-Phe-Ala) was synthesized using the methodology and assessed for yield and purity. The results showed that organic solvent-based ultrafiltration can be a helpful tool for peptide synthesis in green solvents.

Patterned Membranes for Biofouling Mitigation by Electrospinning

Amos Taiswa (Montana Tech)*

Biofilm buildup on water filtration membranes impedes flow, increases energy consumption, and diminishes filtration efficacy resulting in public health concerns. Recently, antifouling technology development has steered away from chemical means that limit use or create toxicity concerns. One such non-chemical method for reducing biofilm includes using biomimicry to pattern filter designs that reduce biofilm through physical means found in nature. Biomimetic designs can foster properties such as enhanced surface wetting that improves fouling resistance without eliciting a damaging environmental impact. In this study, electrospinning was used to deposit polymeric antifouling patterns on mixed cellulose ester membranes for improved filter performance. Aligned polycaprolactone fibers deposited onto the membranes that contained a 300–500 nm diameters and 1–10 μm square grid spacings provided sufficient bacterial disturbance that resulted in a >30 % reduction in biofilm biovolume in both static and dead-end filtration processes. Further, these fibers exhibited resistance to wettability with contact angles of 102°. Despite added fibers that reduced flow through the membrane, filtration performance remained comparable to the controls. Herein, this work provides valuable insight into how electrospinning can be used to provide environmentally benign nanopatterns that improve filtration performance.

Investigating the electric field driven membrane separation of biomolecules towards increased viral vector purity

Emily Ingram (University of Kentucky)*; Trupte Patel (University of Kentucky); Mara Leach (University of Kentucky); Keshav Bhanderi (University of Kentucky); Kearstin Edmonds (University of Kentucky); Rebecca Dutch (University of Kentucky); Malgorzata Chwatko (University of Kentucky)

Viral vectors are modified viruses that can deliver genetic material to cells to cure genetic diseases such as sickle cell anemia and cancer. Ensuring safety and efficacy, separation technology is key to purifying viral vectors from various sized contaminants generated during synthesis. Additionally, biomolecules such as proteins, DNA, and viral vectors possess charges that could migrate in response to an applied electric field. Charge-based separation via electrodialysis is widely used in water desalination with recent advancements enhancing efficiency. Yet, its application in bioprocessing is underexplored. Inspired by electrodialysis, we propose employing membrane-assisted electro-separation for size and charge based separation of biomolecules. In this study, we evaluated the impact of voltage potential, electrolyte solution composition, and membrane pore size on biomolecule permeation. Using a static multi-compartment membrane cell, separated by semi-permeable membranes, suspended in an electrolyte solution, pore size impacts were tested. Bovine serum albumin (BSA), 20 nm and 200 nm silica nanoparticles (SNPs), served as representative host cell proteins, viral vectors and host cell debris, respectively. Higher voltages and longer operational times shifted system pH, affecting biomolecule migration. Increased buffer capacity mitigated pH drops below the biomolecules' isoelectric point. With optimized conditions, increasing pore size from 0.22 μm to 0.45 μm increased BSA permeation from 40% to 72%, attributed to protein aggregation during filtration. Shifting starting pH in a mixed protein-particle model from 7 to 4, increased BSA's net positive charge and maintained SNPs negative charge. This led to full retention of 200 nm SNPs, with 19% and 7% permeation of 20 nm SNPs and BSA, respectively. Integration of membrane-assisted electric field-driven separation offers a tunable method for biomolecule separation, specifically increasing viral vector purity.

Biocompatible Polyacrylonitrile Hollow Fiber Membranes with Enhanced Hemodialysis Performance

Pranav Krishna (Indian Institute of Science Education and Research (IISER) Bhopal India);
Debasis Nayak (Indian Institute of Science Education and Research (IISER) Bhopal India);
Akshay Modi (Indian Institute of Science Education and Research (IISER) Bhopal India)*

In the last few decades, the rising global burden of end-stage renal disease has significantly increased the demand for effective hemodialysis treatments. Concurrently, advancements in membrane technology have enabled the development of more efficient and biocompatible dialyzer membranes (Pavlenko et al., 2016). However, achieving high flux and biocompatibility remains a critical challenge in hemodialysis. This study investigates polyethylene glycol (Mw = 1000; PEG-1000) as a low-cost, safe, and environmentally benign additive to fabricate high-performing functional polyacrylonitrile (PAN) hollow fiber membranes under optimized conditions for hemodialysis applications. Various concentrations of PEG-1000 were utilized, which resulted in desired physicochemical properties. The surface morphology of PAN membranes underwent a significant transformation with PEG-1000 incorporation, shifting from finger-like to spongy structures, which enhanced the clearance of uremic toxins (~60%). The modified membranes showed pure water permeability of 296.5 mL/m^2/h/mmHg and demonstrated >98% protein retention, outperforming pristine PAN membrane. Biocompatibility assessments revealed enhanced cell viability and negligible hemolysis (<1%). Interestingly, no adverse post-surgery reactions were observed when surgically implanted in small rodent models. The animals remained healthy, and retrieved implants showed no signs of fibrosis or inflammation after 21 days, confirming the excellent biocompatibility of the membranes. These findings highlight the potential of PEG-blended PAN membranes for improved hemodialysis performance.

Reference:
Pavlenko, D., Van Geffen, E., Van Steenbergen, M. J., Glorieux, G., Vanholder, R., Gerritsen, K. G., & Stamatialis, D. (2016). New low-flux mixed matrix membranes that offer superior removal of protein-bound toxins from human plasma. Scientific Reports, 6(1), 34429.

Understanding the role of fixed charged group hydrophobicity on nitrate selectivity in anion-exchange membranes

Harsh Patel (University of Michigan)*; Alexander Bentley (University of Michigan); Jovan Kamcev (University of Michigan)

The widespread use of nitrate-based fertilizers to enhance food security has led to significant nitrate pollution in natural water sources. This contamination presents health risks, as nitrates can convert to nitrites, impairing oxygen transport in the bloodstream. Ion-exchange membranes (IEMs), which are polymeric membranes containing charged functional groups covalently bound to their backbone (i.e., fixed charges), are used in electrochemical ion recovery applications. However, most commercial IEMs do not exhibit high nitrate selectivity over competing anions such as Cl– since the main mechanism for separation is electrostatic exclusion. Previous studies have explored enhancing nitrate selectivity by modifying the membrane fixed charged group (FCG) hydrophobicity to leverage differences in ion hydration energy. However, the influence of FCG hydrophobicity on nitrate selectivity remains unclear, as other membrane properties that may impact selectivity (i.e., charge and water content) are not independently controlled in these studies. In this study, we synthesized a series of charged monomers with varying FCG hydrophobicity and fabricated cross-linked anion-exchange membranes (AEMs) with constant charge density and water content via free radical polymerization to isolate the effect of FCG hydrophobicity. We measured the equilibrium counter-ion partitioning selectivity of the AEMs with Cl-/NO3- and NO3-/SO42- salt mixtures across a wide range of concentrations and equivalence fractions. We also measured the counter-ion diffusion selectivity via electrochemical impedance spectroscopy to determine the total transport selectivity. The results demonstrate that increasing FCG hydrophobicity significantly increases NO3-/Cl- and NO3-/SO42- selectivity, suggesting that specific ion effects, which are intensified in hydrophobic environments, play a critical role in improving nitrate selectivity.

High-throughput ion permeability analysis of 2D laminar membranes using robotic ICP platform

Min A Kim (Argonne National Laboratory)*; Ilya Shkrob (Argonne National Laboratory); Yining Liu (University of Chicago); Seth Darling (Argonne National Laboratory)

Phyllosilicates have emerged as abundant and economical source materials for membranes featuring laminar structures and tunable physicochemical properties providing unique ion sieving capabilities. To rapidly explore the complex structure-chemistry-function-performance space of phyllosilicate membranes, spanning different minerals, crosslinkers, dopants, functionalization, and thickness, the development of a high-throughput (HT) autonomous experimental platform is essential. Herein, we demonstrate a streamlined high-throughput (HT) workflow for ion permeability screening. Using a mobile manipulator for sample transport and equipment manipulation, an automated screening process integrates batch diffusion experiments with their permeate analysis. We implemented a liquid-handling robotic platform to expose phyllosilicate membranes to a library of ions in parallel. The permeate solution was analyzed by HT ICP to observe changes in ion concentration over time. We used phyllosilicate membrane as a model system; our modular robotic workflow can be easily adapted to screen other membranes. The HT workflow provides an efficient way to tackle repetitive and time-consuming tasks with minimal human intervention, allowing characterization of membrane ion permeability in studies that would be intractable via traditional manual operation and compatibility with AI/ML integration.

Role of Frictional Forces in Salt and Water Transport through Polyamide Reverse Osmosis Membranes

Kevin Pataroque (Yale University)*; Menachem Elimelech (Rice University); Subhamoy Mahajan (University of Wisconsin Madison); Martina Del Cerro (Yale University); Jonathan Mejia (Yale University); Hanqing Fan (Rice University)

Transport models can be used to better understand water and salt transport in conventional polyamide reverse osmosis (RO) membranes, enabling further development of RO technology. However, current models do not capture the full extent of interactions between the water molecules and ions with the membrane matrix. In this study, we establish relationships between intrinsic membrane properties and transport parameters based on the solution-friction model. Nominal pore radii of various commercial membranes are characterized by measuring the rejection of neutral organic solutes. Salt partitioning is quantified with quartz crystal microbalance (QCM), while salt permeance is measured in an RO dead-end cell. While salt partitioning increased by 34% as nominal pore size increased from 0.30 to 0.35 nm, salt permeance increased more than tenfold across this range. The difference in magnitude shows that salt permeance behavior cannot be fully attributed to salt partitioning. Using the solution-friction model, which considers friction between species (ions, water, and the membrane matrix) to predict transport behavior, we further show how trends in salt and water permeance can be captured by ion-membrane and water-membrane frictional coefficients across various pore radii, showing that frictional interactions control transport through the membrane. These findings are expanded upon using non-equilibrium MD simulations, which correlate these frictional coefficients to ion and water self-diffusivity coefficients. In membranes with higher fractional free volume, frictional interactions between these species and the membrane matrix decrease, raising the self-diffusivity of these species to increase. These results elucidate the role of frictional coefficients in the solution-friction model to physical membrane properties, while demonstrating the influence of these frictional coefficients on other parameters widely used in membrane science.

Carbon Capture and Light Paraffin Separation Membranes Based on Poly(1,3-Dioxolane)/[hmim][Tf2N] Blends

Justin Rosenthal (The University of Texas at Austin)*; Xavier Gatica (The University of Texas at Austin); Antoine Chamoun-Farah (The University of Texas at Austin); Adam Mann (The University of Texas at Austin); Tangqiumei Song (The University of Texas at Austin); Isaac Tan (The University of Texas at Austin); Enrique Loredo (The University of Texas at Austin); Nathaniel Lynd (The University of Texas at Austin); Gabriel Sanoja (The University of Texas at Austin); Joan Brennecke (The University of Texas at Austin); Benny Freeman (The University of Texas at Austin)

Global climate change driven by anthropogenic carbon dioxide (CO2) emissions is one of the most urgent challenges facing humanity. As part of efforts to mitigate this issue, there is a growing demand for more energy efficient technologies for gas separations. Membrane-based separation technologies, which require significantly less energy than traditional thermal processes, offer a promising solution. Gas separation membranes have great potential for selectively separating CO2 from other gases such as nitrogen (N2) and methane (CH4) for sequestration or utilization. Additionally, they can enhance the energy efficiency of other separations, such as C2+ hydrocarbons from CH4, thus further reducing CO2 emissions. Polyether membranes based on poly(1,3-dioxolane) (PDXL) have shown desirable CO2/N2 separation performance due to their polar ether oxygen groups which exhibit a strong affinity toward CO2. Furthermore, PDXL is compatible with a variety of ionic liquids (ILs), enabling the tailoring of its separation capabilities. In this study, a variety of PDXL/IL blend membranes were synthesized to improve gas separations for both CO2 and light paraffins. The resulting membranes were evaluated for their pure- and mixed-gas transport properties. We found that the addition of IL significantly enhanced gas permeabilities (e.g., CO2 increased from 330 Barrer to 600 Barrer), and also influenced the resulting permselectivity. This work demonstrates that PDXL/IL blends offer highly permeable and tunable membranes suitable for multiple gas separation applications depending on the choice of IL.

Engineering Cobalt Selective Membranes to Support the Energy Transition

Lauren Mazurowski (Yale University)*; Zhongren Jiao (Yale University); Menachem Elimelech (Rice University)

Global decarbonization relies heavily on cobalt, a critical component in the manufacture of lithium-ion batteries and large-scale energy storage. Cobalt-rich industrial wastewaters and natural brines exhibit great potential to augment cobalt supply, but current recovery methods from aqueous sources remain costly and chemically intensive. Membranes present a promising alternative as a chemical-free technology to replace organic solvent-heavy processes; however, they often lack the precision needed to separate cobalt from other relevant competing species, such as nickel, lithium, and sodium. In this study, we aim to develop a cobalt-selective membrane in which selective transport of cobalt is facilitated by the preferential interaction of cobalt with phosphonic acid functional groups. The membrane is fabricated by alternately layering polyallylamine hydrochloride (PAH) and a novel phosphonic acid polymer (PPA), the latter synthesized in-house. Upon successful layer-by-layer deposition of these polyelectrolytes onto an anodic aluminum oxide substrate, the membranes are evaluated for selectivity using various single and multi-salt solutions in diffusion dialysis. Our initial findings have demonstrated lithium/cobalt selectivity of nearly 50, and cobalt/nickel selectivity exceeding 2—a particularly challenging separation that traditionally depends on heavy organic solvent use.

Based on the promising selectivities obtained with the membranes in this study, future work will focus on the incorporation of such membranes into an electrodialysis configuration to increase ion throughput.

Harnessing Onsite Water Reuse for Sustainable Data Center Operations

Reema Shinh (University of Arizona)*; Leila Karimi (University of Arizona); Kerri Hickenbottom (University of Arizona)

Data centers (DCs) have been growing the adoption of the internet, artificial intelligence (AI), streaming, e-commerce, and cloud computing, encouraging data expansion over the last century while simultaneously driving a greater water footprint. Nearly 57% of water consumed in DCs is potable [1] – a stressor for utility companies that aim to meet the demands of residents, commercial businesses, and industries. ChatGPT alone, an open AI language model platform, consumes nearly 700,000 liters of water for training alone because of cooling the systems [2]. Though energy efficiencies have been extensively researched and optimized, some gaps remain in the literature surrounding solutions for meeting growing water demands in DCs. Transitioning to alternative water sources for DC cooling has the potential to create significant strides for water reuse adoption while also having an economic incentive to increase water efficiency. Alternative water sources such as harvested rainwater, cooling tower blowdown, municipal reclaimed water, and captured condensate are promising alternatives for non-potable reuse. The objective of this study was to enhance DC sustainability by addressing potential onsite water reuse opportunities to reduce water consumption by (1) evaluating constituents and availability of alternative water sources, (2) modeling membrane-based treatment technology to the water source, and (3) assessing the techno-economic feasibility of each design configuration specific to arid regions. This study further investigates key parameters that largely influence the levelized cost of water (LCOW) and the net present value through a series of sensitivity analyses. Results highlighted in this project indicate that harvested rainwater and captured condensate as alternative water sources provided the highest water cost savings of up to \$0.48 m-3 and the lowest payback period of nearly 2.6 years for a 34.5 MW DC case study.

Domain Aware Modeling of Multi-Mode Operation of RO water treatment and desalination system via Graph Convolutional Networks

Michael Clement (California State University, San Bernardino); Nathan Kelly (California State University, San Bernardino); Nora Marki (University of California, Los Angeles); Maria Soto (University of California, Los Angeles); Yoram Cohen (Professor); Bilal Khan (California State University, San BErnardino)*

Distributed water treatment and desalination systems are essential for upgrading impaired inland groundwater of remote communities. However, given the lack of onsite 24/7 technical operational oversight, it is critical for such systems to autonomously adapt to water demand and temporally variable source water quality. To achieve the above, machine learning (ML) models are needed to describe and forecast system performance, provide a basis for model predictive control (MPC), detect operational and sensor faults, and identify and quantify causal relationships among process attributes. Accordingly, a modeling approach was developed, based on graph convolutional neural networks (GCNN) with attention mechanism (GCNN-AT), to describe the intermittent operational profiles of spiral-wound RO system. Models for permeate flux and solute passage predictions were developed based on wellhead spiral-wound RO system operation (deployed in Salinas Valley, CA) capable of permeate production up to 4,500 gallons/day. Time-series data from 22 system sensors (pressure, flow rate, conductivity, temperature of the feed, concentrate and permeate streams) was compiled over four operational modes (startup, permeate production, shutdown, flushing) during a year of operation yielding over 6 million data samples. The GCNN-AT models, constructed with 3 hidden layers, 16 hidden nodes and 10 attention heads per layer, demonstrated excellent forecasting prediction (1 month forward in time) with $R^2 > 0.95$, suitable for identifying operational faults, membrane performance degradation. Causal relationships were extracted from the models by aggregating and visualizing the learned weights within the graph neural network structure. The analysis revealed that temperature had a significant impact on permeate flux and solute passage, in addition to the expected dependence on feed salinity and nitrate levels, as well as concentrate recycle ratio, and flow rate and pressure profile along the RO elements.

Hyperselective carbon molecular sieve membranes for precise H2 and CO2 separation

Gaurav Iyer (University of Maryland); Ching-En Ku (University of Maryland); Chen Zhang (University of Maryland)*

Hydrogen (H2) has emerged as a clean and sustainable fuel for energy generation and transportation. Globally, over 90% of H2 production is sourced from fossil fuels, which requires energy-intensive downstream purification to recover high-purity H2 and capture the carbon dioxide (CO2) byproduct. While membranes can decarbonize H2/CO2 separation, their moderate H2/CO2 selectivity necessitates secondary H2 purification by pressure-swing adsorption. Membranes with ultra-high H2/CO2 selectivity will enable one-step enrichment of fuel-grade (>99.9%) H2 without secondary purification.

In this presentation, we will report a novel hyperselective carbon molecular sieve (CMS) membrane with ultra-high H2/CO2 selectivity [1]. The hyperselective CMS membrane demonstrates H2/CO2 selectivity over 7000 under mixed-gas permeation at 150°C, which is almost 30 times higher than the most selective non-metallic membranes reported in literature and significantly exceeds the polymer upper bound. Pore structure characterization suggests that highly refined CMS ultramicropores are responsible for precise H2 and CO2 separation. Modeling shows that the hyperselective CMS membrane will enable one-step enrichment of 99.99% pure H2 from shifted syngas with high H2 recovery. The findings of this work thus pave the way for simultaneous H2 purification and CO2 capture for decarbonized H2 production.

References:

[1] Iyer, G.; Ku, C.; Zhang, C., Hyperselective carbon membranes for precise high-temperature H2 and CO2 separation, under revision 2025

Dynamic Shifts in Gas Transport: A Framework for Discriminating Solution-Diffusion and Pore-Flow Mechanisms via Media Atom Motion

Jianhao Qian (Rice university)*; Ruoyu Wang (Rice university); Menachem Elimelech (Rice university)

Gas transport is a ubiquitous phenomenon in both natural and industrial contexts, commonly categorized as pore-flow or solution-diffusion mechanism. These mechanisms exhibit significant theoretical distinctions, making the effective discrimination between them crucial for accurate prediction of gas transport and separation processes. However, the identification of these transport mechanisms remains largely subjective, lacking quantifiable and efficient criteria. Herein, we show that the fundamental difference between the two mechanisms lies in whether gas transport is determined by the motion of media atoms. In the solution-diffusion mechanism, media atoms diffuse and carry the gas molecules along, whereas in the pore-flow mechanism, media atoms only alter the direction of gas molecules through collisions or adsorption. We introduce, for the first time, a dimensionless number that represents the motion ratio of media atoms to gas atoms, which effectively distinguishes between solution-diffusion and pore-flow mechanism. This dimensionless number, combined with the Knudsen number, provides a systematic framework for classifying gas transport mechanisms. Based on this proposed dimensionless number, we predict that some traditional solution-diffusion processes may transition to pore flow as the motion of media atoms slows down. Using molecular dynamics simulations, we investigated gas diffusion through a water film and observed that the gas transport mechanism can shift from solution-diffusion to pore-flow due to the freezing of the water medium. These findings underscore the dynamic nature of gas transport mechanisms, emphasize the critical role of media atom motion in the analysis of gas transport processes, and provide a robust tool for effectively distinguishing between gas transport mechanisms.

Designed for Recycling: Dynamic Crosslinking of Solvent-Resistant Membranes

Malinalli Ramirez Martinez (King Abdullah University of Science and Technology)*; Sandra Aristizabal (King Abdullah University of Science and Technology); Lakshmeesha Upadhyaya (King Abdullah University of Science and Technology); Abdul-Hamid Emwas (King Abdullah University of Science and Technology); Nikos Hadjichristidis (King Abdullah University of Science and Technology); Suzana Nunes (King Abdullah University of Science and Technology)

Membrane recycling is a way to improve the sustainability of the membrane industry, promoting circularity and lowering the environmental impact associated with the membranes' end-of-life. However, recycling is challenging for crosslinked polymer membranes, which are required for applications demanding high chemical stability, such as the filtration of organic solvents. To tackle this challenge, we propose a straightforward reversible crosslinking approach for Ultem® membranes, in contrast to the traditional permanent covalent crosslinking methods.

The reversible crosslinking of Ultem® membranes in flat-sheet and hollow fiber configurations was achieved via post-treatment with cysteamine, which reacted with the polymer in one end and formed a disulfide bond in the opposite end as a result of the thiol groups oxidation. Crosslinked membranes exhibited good stability in organic solvents, such as DMF, NMP, acetone, and acetonitrile, while uncrosslinked Ultem® membranes were unstable in the same solvents. Crosslinking also had a positive effect on the membranes' performance, improving their rejection for Direct Red 80 (MW 1373 g mol–1) in methanol from 75% to 99%. After the filtration tests, the crosslinked membranes were successfully recycled into new ones by reduction of the

disulfide bond with 1,4-Dithiothreitol, which allowed for polymer solubilization in DMF, dope solution preparation, and membrane casting. The recycled membranes exhibited good mechanical stability and a rejection of 85% for Direct Red 80 in methanol, and it was demonstrated that thiol groups remained in the polymer after the recycling process.
This work serves as a demonstration of the great potential of integrating dynamic chemistry and polymer membranes, making it possible to improve their performance while unlocking their recyclability at the end-of-life, thus advancing the efforts towards a more sustainable membrane industry.

Bipolar membranes electrodialysis for water purification and resource recovery

Bin Yang (National Renewable Energy Laboratory)*; Caleb Reese (National Renewable Energy Laboratory); Nurshaun Sreedhar (National Renewable Energy Laboratory;Department of Civil, Architectural and Environmental Engineering, The University of Texas at Austin); Chance Parrish (National Renewable Energy Laboratory); Mou Paul (National Renewable Energy Laboratory); Manish Kumar (Department of Civil, Architectural and Environmental Engineering, The University of Texas at Austin); Abhishek Roy (National Renewable Energy Laboratory)

Reverse osmosis concentrate (ROC) from inland brackish water desalination and wastewater recycling presents moderate salinity, posing significant challenges for treatment and disposal. Existing treatment technologies predominantly rely on energy-intensive thermal methods, large-area evaporation strategies, or deep-well disposal, which often result in the loss of water value. Electrified membrane brine treatment offers a promising alternative by utilizing renewable energy sources, enabling complete water recycling. This work proposes the development of next-generation bipolar membrane electrodialysis (BMED) systems capable of desalinating brine while simultaneously recovering valuable resources. However, several critical challenges remain in the areas of membrane engineering for BMED. These include: 1) decreased current efficiency during desalination due to the low conductivity of feed chambers, 2) fouling and scaling from multivalent ions and contaminants in the feed, and 3) the retention and carryover of ROC contaminants, including PFAS compounds. Our research focuses on the development of novel anion exchange membranes (AEMs) and cation exchange membranes (CEMs) based on synthesized multiblock copolymers. We investigate the effects of block length, ion exchange capacity (IEC), and polymer backbone structure on cation and anion transport (e.g., Na^+, Cl^-, OH^-, H^+) for both our synthesized membranes and commercial membranes. Fundamental properties of these membranes, such as water uptake, IEC, and ionic conductivity, are also evaluated to optimize performance for BMED applications.

Purifying circular RNA by ultrafiltration

Karen Guillen Cuevas (Clemson University); Xiaoming Lu (Clemson University); Marc Birtwistle (Clemson University); Scott Husson (Clemson University)*

In vitro synthesized circular RNA (circRNA) is promising for developing RNA therapeutics due to its high stability and low immunogenicity compared to linear RNA. However, commercial-scale production remains challenging due to the lack of efficient and scalable purification methods. The immunogenicity and protein expression stability of circRNA depend strongly on purity, but existing purification techniques have significant limitations: gel electrophoresis is not scalable, exonuclease degradation is costly with low yields, and size exclusion high-performance liquid chromatography (SE-HPLC) suffers from peak overlap, resulting in low yield and selectivity. While combining SE-HPLC and RNase R enrichment achieves 90% purity, it does so at the expense of yield (<50%), underscoring the pressing need for innovative, scalable, high-yield purification strategies.

We assessed ultrafiltration/diafiltration (UF/DF) for purifying protein encoding circRNA. We determined separation conditions by measuring the sieving coefficients of purified circRNA and linear RNA conformers as a function of permeate flux. Using suitable separation conditions, we measured purity and yield for purifying circRNA from mixtures of these RNA molecules. For a ternary mixture of circRNA, precursor RNA, and nicked RNA, UF/DF achieved 96% circRNA purity with 53% yield. When applied to diluted in vitro transcription (IVT) products, UF/DF achieved 86% purity with >50% yield, significantly outperforming SE-HPLC, which achieved only 41% purity with 45% yield for the same IVT samples. Additionally, we evaluated the cytotoxicity of UF-purified circRNA using reverse transfection in HEK293T cells. Our findings establish UF/DF as a superior research-scale method for purifying circRNA and suggest its strong potential for large-scale manufacturing of circRNA therapeutics.

Utilizing a Green Solvent for Energy-Efficient CO2 Capture from Industrial Emission Sources via Hollow Fiber Membrane Contactors

Nicholas Gregorich (Oak Ridge National Laboratory)*; Mary Irwin (Oak Ridge National Laboratory); Zachary Coin (Oak Ridge National Laboratory); Stephen DeWitt (Arkema Inc); Ramesh Bhave (Oak Ridge National Laboratory); David Sholl (Oak Ridge National Laboratory); Syed Islam (Oak Ridge National Laboratory)

As human activities continue to release more greenhouse gases than natural processes can remove, the challenge of reversing the global warming crisis remains at the forefront of modern research. Amine-based processes are the most mature technology in the carbon capture space but require high energy during the regeneration process and release degraded, carcinogenic emissions that require further processing. To overcome these challenges, absorption of carbon dioxide (CO2) via environmentally friendly, green solvents, has emerged as a potential substitute for state-of-the-art amines. The use of diethyl sebacate (DS) as a solvent for pre- and post-combustion carbon capture presents many advantages that include high CO2 solubility, high hydrophobicity, low viscosity, and high CO2 selectivity over N2. When considering DS is a commercially available green solvent commonly used in food and cosmetics industries, scaling up of this process will not be hindered by the supply chain. This study demonstrated that DS captures CO2 at industry-grade purity (>95%) from a CO2/N2 mixture in a fully continuous process that requires zero external energy for solvent regeneration. This process utilizes low-cost, scalable polymer hollow fiber membrane contactors (HFMC) that require low solvent inventory and provide high interfacial area for energy-efficient CO2 capture from simulated flue gas streams. Technoeconomic analysis of this process was performed at a scale comparable to active natural gas power plants with emissions containing 10% CO2. Scaling this process to 2-million tons of CO2 annually requires a 5-stage process that operates at a processing cost of ~$93 per ton of CO2 captured. Results from this work underscore the importance of integrating green solvents with proven membrane technologies, could alleviate the burdens of amine-based capture, and holds strong potential for industrial deployment in the growing carbon capture ecosystem.

Permeability and Selectivity Model for Separation Membranes Based on Fractal Analysis of Porous Structure

Riccardo Lovison (University of Colorado Boulder)*; Yifu Ding (University of Colorado Boulder); Christina Carbrello (MilliporeSigma)

Accurate characterization of intricate 3D porous membrane structures is essential for the development of advanced membranes. Specifically, correlating membrane pore structure with permselective properties remains a critical challenge for membrane technologies. Typically, pore size distribution (PSD) is routinely quantified with liquid-liquid and gas-liquid displacement porosimetry, using Young-Laplace equation. The challenges of these techniques involve the accurate estimation of maximum and/or minimum pore size., Often the pore size determined from these methods do not correlate with SEM images of the membranes.

Currently, there are no methods predicting permselective properties completely using real-space images. This study introduces a novel image-based methodology for characterizing the porous structure of microfiltration (MF) and ultrafiltration (UF) membranes using fractal analysis. The core of the methodology involves extracting two key fractal dimensions: pore-area fractal dimension (D_f), obtained from the in-plane cross-sectional images of the membrane, and tortuosity fractal dimension (D_T) obtained from in-thickness cross-sectional images.

Combining porosity, maximum pore size, thickness and fractal dimensions (D_F and D_T), a Fractal Permeability Model (FPM) and a Fractal Selectivity Model (FSM) have been developed to predict permselective properties of the membranes. The FPM has been verified first using symmetric PVDF membranes. Then, the model has been applied to asymmetric PES membranes, using a multi-layer subdivision of the structure, demonstrating good agreement with experimental results. The FSM accurately predicted typical log reduction values (LRV) for virus filtration membranes, showing good agreement with literature reports on separation of parvoviruses and bacteriophage.

Specific Ion Selectivity Between Monovalent Cations in Commercial Cation Exchange Membranes

Vaishnavi Koyilath Nandakumar (Columbia University)*; Ngai Yin Yip (Columbia University)

As the demand for more efficient and sustainable separation technologies grows, understanding specific ion selectivity in ion exchange membranes (IEM) becomes increasingly important. Specific ion selectivity between monovalent cations remains a relatively unexplored field of study. This selectivity plays a critical role in the performance of processes such as electrodialysis ion separations, energy storage, and material recovery from waste streams, where the preferential transport of an ion over others of like valency is essential. This study aims to address this gap by investigating the preferential sorption, mobility, and overall transport between monovalent cations through IEM. Eleven commercially available cation exchange membranes, each with distinct chemistry, structure, and morphology were assessed in binary cation systems (Li+/K+), i.e., mixed electrolytes. All eleven membranes exhibit specific ion selectivity between the two monovalent cations for sorption, mobility, and overall transport, suggesting ion-membrane interactions that are not fully captured in the current understanding. Sorption demonstrated a clear preference for K+ while mobility selectivity varied, with the majority of membranes showing selectivity for Li+ and a few exhibiting exceptions. Combining the sorption and mobility selectivity yields the overall transport selectivity. Interestingly, despite the notable differences in specific ion preferences observed for sorption and mobility between Li+ and K+, the overall transport selectivity of most IEMs closely align with those observed in bulk solution. By shedding light on the underlying mechanisms influencing specific ion selectivity between monovalent cations in IEM, the study aims to contribute to the development of advanced, energy-efficient IEM for future applications, enabling precise control of ion transport.

Flat Sheet Membrane Modules for Gas Separation to Enhance Countercurrent Contacting

Hamed Rahnema (University of Toledo); Kale Harbick (USDA); Glenn Lipscomb (University of Toledo, Toledo)*

High-performance membranes are commonly fabricated as hollow fibers or flat sheets. Hollow fiber modules inherently allow counter-current flow, while standard flat-sheet modules, plate-and-frame or spiral-wound, typically rely on crossflow. As a result, additional design features are required to enable counter-current contacting in flat-sheet modules, and the efficacy of these features in achieving such flow without imposing excessive pressure drops remains poorly understood.

In flat sheet modules, glue lines are often added to direct fluid flow anti-parallel to the retentate flow (i.e., to introduce partial counter-current contacting). Nevertheless, a degree of crossflow is still needed to distribute fluid along the entire leaf. Glue lines also negatively impact membrane performance in two ways: (1) by reducing the membrane area available for permeation, and (2) by introducing extra pressure drop.

This study examines four distinct glue-line configurations, used to modify spiral wound modules, disclosed in the patent literature: (1) multiple glue lines running the full length of the membrane leaf, (2) variations in glue-line shape, (3) changes in glue-line thickness, and (4) placement of the glue line on either the feed side or the permeate side. These configurations are evaluated using COMSOL Multiphysics simulations for O_2/N_2 separation, comparing module stage cuts, feed flow rates, and pressure drops across for a broad range of nitrogen product compositions. The results are presented for both sweep and non-sweep scenarios.

At lower capture rates, the performance differences between configurations are small. However, as capture rates increase, larger discrepancies emerge, driven by imperfect counter-current contacting, additional pressure drop, and lost membrane area. By quantifying these competing factors, this work provides strategies for effectively incorporating glue lines into flat sheet module designs to balance performance and operational penalties.

Machine Learning–Assisted Inverse Design and Operability Analysis of Membranes for Direct Air Capture

Vitor Renan Vitor Gama (West Virginia University)*; Oishi Sanyal (West Virginia University); Fernando V. Lima (West Virginia University)

Inverse design frameworks—integrating advanced process modeling with machine learning—have emerged as powerful tools that can expedite the research and development of technologies across many scientific fields. In this work, we leverage such an approach to design a multistage membrane-based direct air capture (m-DAC) system employing hollow-fiber modules and novel facilitated transport membrane intrinsic properties. The process simulation, conducted in AVEVA Process Simulation platform and coupled with the Python-based Opyrability library, systematically evaluates the effects of membrane intrinsic properties on key performance metrics, such as: CO2 recovery, purity, and energy consumption (kWh/kgCO_2). Building on a comprehensive two-stage membrane framework, an inverse design study is performed to identify the optimal membrane parameters specifically tuned for DAC conditions. Machine learning models are employed to correlate these parameters with process outcomes, resulting in surrogate models trained for the specific scenarios investigated in this work. Results collected showed that CO2 permeances around 7000 GPU are sufficient to collect a CO2 permeate with concentrations higher than 100 times the atmospheric conditions. Findings from the simulations that consider a rigorous facilitated transport membrane model indicate that a single-stage setup can outperform the simplified process, without sacrificing permeate enrichment. The framework proposed in this research demonstrates how combining process simulations and systems analysis with machine learning can advance m-DAC studies. By directing experimental efforts toward suitable DAC operating regions, this approach expedites discoveries and facilitates the exploration of material properties.

Keywords: Surrogate Modeling, Membrane Simulation, Intrinsic Properties

Solvent Resistance and Adsorption Mechanisms Studies of Deep Eutectic Solvents on Polyvinylidene Fluoride Membranes

Odianosen Ewah (University of Kentucky)*; Isabel Escobar (University of Kentucky)

The urgent need to mitigate climate change, as emphasized by the United Nations Sustainable Development Goal 13, has driven research into reducing CO_2 emissions across industrial sectors, particularly those processing lignocellulosic biomass. As shown in Figure 1, industries such as bioethanol production and pulp and paper manufacturing extract lignin using conventional chemical methods, which are energy-intensive and emit significant amounts of CO_2. Deep Eutectic Solvents (DESs) have emerged as green alternatives for lignin extraction, offering advantages such as low toxicity, biodegradability, and energy-efficient processing. However, the successful implementation of DES-based technologies requires robust separation methods for recovering lignin from DESs. This challenge is compounded by the solvent's potential to degrade many polymeric materials, necessitating the development of solvent-resistant membranes. This study investigates the interaction mechanisms of DESs with Polyvinylidene Fluoride (PVDF) membranes, chosen after polymer screening using Relative Energy Difference (RED) calculations and supported by experimental validation to ensure solvent resistance. PVDF membranes were subjected to adsorption experiments to evaluate DES-membrane interactions. The adsorption kinetics were modeled using pseudo-first-order and pseudo-second-order equations, while isotherm behavior was analyzed through Langmuir and Freundlich models to elucidate equilibrium dynamics. Preliminary findings underscore the importance of polymer compatibility with DESs in preserving membrane integrity and performance. The adsorption studies aim to identify optimal membrane conditions for the efficient separation of DESs from lignin, enabling energy-efficient solvent recovery without compromising membrane durability. This research contributes to the broader effort of reducing CO_2 emissions by advancing DES-based technologies for sustainable lignin extraction and processing.

Simplified Synthesis of Biomimetic Self-Assemblies for Nanofiltration and Ion-Ion Separations

Ronald Vogler (The University of Texas at Austin)*; Katherine Kimball (The University of Texas at Austin); Aída Fica (The University of Texas at Austin); Dominic Bujanos (The University of Texas at Austin); Benny Freeman (The University of Texas at Austin); Manish Kumar (The University of Texas at Austin)

The authors chose not to publish the abstract content.

A channel based paradigm for membrane separations

Manish Kumar (University of Texas at Austin)*

Fast and selective water and ion transport across cellular membranes has inspired the development of bioinspired membranes for possible separations applications . An emerging approach is to use artificial channels (ACs) as a key transport element of membranes for aqueous separations such as desalination and ion-ion separations. ACs are synthetic mimics or engineered versions of naturally occurring biological membrane proteins. In this talk, I will provide a brief overview of recent and exciting AC developments with a focus on artificial water and ion channels and discuss emerging paradigms for membrane architectures that can utilize these channels. I will also provide a perspective on how these novel artificial water and ion channels could be leveraged to create high performance membranes using some proposed general principles for channel-based membranes.

Interfacial polymerization of micropore-generating units for molecular separation: Beyond desalination

Taehoon Lee (Sungkyunkwan University)*; Zachary P. Smith (Massachusetts Institute of Technology)

Interfacial polymerization has been an industrial standard for preparing desalination membranes. Extending the same concept to nonaqueous environments (i.e., organic solvent nanofiltration/reverse osmosis, OSN/OSRO) would be a key enabler for the decarbonization of the chemical and petrochemical industries. However, a classical, fully aromatic polyamide thin-film composite (TFC) membrane often fails to present desired separation ability of nonaqueous mixtures due to its hydrophilic nature, low free volume elements, and poor swelling/plasticization resistance. To overcome this issue, we herein report a molecular engineering approach based on "acid-catalyzed interfacial polymerization" to devleop advanced TFC membranes for efficient molecular separation in organic solvents. The design strategies include (1) alternating the linkage from amide to imine and (2) subsequent introduction of shape-persistent and micropore-generating units. The developed membranes feature excellent separation performance for nonaqueous liquid and ion separation for crude oil fractionation and redox flow batteries, respectively. Our results highlight that introducing these moieties is a facile and useful strategy to develop size-selective, plasticization-resistant, and fluorine-free OSN- or OSRO-range membranes by interfacial polymerization.

Linking Biomass Properties and Membrane Fouling in an Anaerobic Membrane Bioreactor for Domestic Wastewater Treatment

Mor Shental-Isaacs (Ben Gurion University of the Negev); Moshe Herzberg (Zuckerberg Institute for Water Research, Blaustein Institutes for Desert Research, Ben-Gurion University of the Negev)*; Ronit Bitton (Department of Chemical Engineering and the Ilse Katz Institute for Nanoscale Science &Technology, Ben-Gurion University of the Negev); Michal Zaiden (Department of Chemical Engineering and the Ilse Katz Institute for Nanoscale Science &Technology, Ben-Gurion University of the Negev)

Anaerobic Membrane Bioreactors (AnMBRs) offer a sustainable solution for domestic wastewater (DWW) treatment by integrating anaerobic digestion and membrane filtration. This approach converts organic matter into methane-rich biogas while reducing energy use, minimizing sludge production, and producing effluent suitable for reuse and nutrient recovery. Despite these advantages, membrane fouling remains a key challenge, limiting operational stability and increasing costs.

This study investigates fouling mechanisms in AnMBRs using innovative biomass characterization techniques. Conducted at the Shafdan wastewater treatment plant in Israel, our research utilized an 8 m^3 AnMBR pilot treating raw DWW. We examined the effects of increased organic loading rates (OLRs) on biomass properties and fouling behavior using state-of-the-art methods, including quartz crystal microbalance with dissipation monitoring (QCMD) coupled with localized surface plasmon resonance (LSPR). This revealed critical insights into extracellular polymeric substances (EPS), their viscoelastic properties, and interactions with membrane surfaces. Dynamic rheometry quantified floc mechanical integrity, while Sludge Volume Index (SVI) and Capillary Suction Time (CST) assessed dewaterability.

Our findings show that higher OLRs alter EPS composition, reduce floc stability, and impair biomass dewaterability. Scanning Electron Microscopy and Energy Dispersive X-ray analyses confirmed reduced calcium levels in destabilized flocs. Increased transmembrane pressure and Optical Coherence Tomography imaging demonstrated the link between biomass changes and fouling propensity.

This research advances understanding of biomass behavior in AnMBRs, providing actionable insights to optimize fouling mitigation strategies. These findings support the development of more energy-efficient, cost-effective, and sustainable wastewater treatment systems.

Silica functionalized membranes from solid waste and its utilization in water splitting and water treatment applications

Pallabi Das (CSIR-Central Institute of Mining and Fuel Research)*; Satyendra Kumar (CSIR-CIMFR); Babli Prasad (CSIR-CIMFR); Srivatsa Bettahalli (University of Connecticut); Jeffrey McCutcheon (University of Connecticut); Suman Dutta (Indian Institute of Technology Indian School of Mines)

Aluminosilicate functionalized membranes with high selectivity, robustness, finds use in ion selective recovery, water purification, and energy production. Their intrinsic molecular sieving properties, high ion-exchange capacity and chemical resilience makes them useful in multifunctional membrane development. Traditionally synthesis of these membranes involves chemical precursors as silica and alumina sources. However, industrial byproducts rich in aluminosilicates such as slag (a waste product from steel production) and fly ash (a byproduct of coal combustion) provides alternative aluminosilicate sources. Structural modification and zeolilization of slag and fly ash in recent years has gained attention due to the cost-effectiveness and the availability of large volumes of such waste.

In this work, we present some of our key findings on silica functionalized membranes for remediation, selective recovery and water splitting applications. Steel slag-derived aluminosilicate membranes synthesized through hydrothermal route demonstrated high surface reactivity and ion-exchange properties. These membranes showed efficient fluoride removal in presence of competing anions like nitrate. Similarly, fly ash-derived zeolite membranes demonstrated over 80% ammonium recovery due to a precise microporous structure and high aluminium content.

In another application titania-doped aluminosilicate membranes were synthesized using the sol-gel route and evaluated for water splitting. Effect of surface functionalization was investigated for enhancing photocatalytic efficiency. Incorporation of titania into the aluminosilicate framework enhanced light absorption, charge separation under UV and visible light irradiation.
This highlights significant potential of slag, fly ash, and titania-functionalized aluminosilicate membranes as alternative feedstocks to conventional precursors paving the way for viable waste utilization and circular economy in this domain.

Influence of the temperature and surface charge on the performance of nanocellulose based facilitated transport membranes for carbon capture

Elena Medri (Chemical and Process Engineering Department - Alma mater studiorum Bologna)*; Marco Giacinti Baschetti (University of Bologna); Ferruccio Doghieri (University of Bologna); Stefano Rossini (ENI s.p.a.); Luca Fascia (ENI s.p.a)

In recent years, significant efforts have been made to develop efficient and cost-effective methods for CO2 separation from light gases with applications in the CCS value chain. Gas separation membranes have emerged as promising candidates, and ongoing research focuses on optimizing their performance. Previous studies demonstrated that nanocellulose-based facilitated transport membranes with L-arginine as a mobile carrier enable reliable CO2/N2 separation for post-combustion applications [1,2]. This study expands on these findings by examining how temperature variations and superficial charge density (SCD) affect gas permeation under humid conditions (70%–90% RH) in three nanocellulose membranes: softwood-derived, hardwood-derived, and carboxymethylated hardwood with a SCD of 780 μequiv/g. Permeation tests were conducted under controlled humidity and different temperatures using a fixed-volume, variable-pressure apparatus. CO2 and N2 permeability, along with selectivity, were analyzed in pristine and L-arginine-modified membranes. Results showed that pristine nanocellulose membranes exhibited increased permeability with rising humidity, while SCD significantly influenced CO2 permeability and selectivity. Pristine hardwood nanocellulose revealed a CO2 permeability of 10 Barrer at 95% RH and room temperature, increasing to 38 Barrer under the same conditions with SCD at 780 μequiv/g, while selectivity rose from 15 to 120. Adding L-arginine further enhanced separation performance: CO2 permeability doubled for softwood nanocellulose, tripled for hardwood nanocellulose, and increased 1.5-fold for 780 μequiv/g nanocellulose at room temperature and 90% RH. A similar trend was observed for selectivity, confirming the effectiveness of facilitated transport membranes in improving CO2/N2 separation. Temperature had a major impact, as increased thermal energy enhanced carrier mobility, improving CO2 transport.

Thiol-Ene Photopolymerization for Highly Selective Anion Exchange Membranes with High Charge Concentration

Jocelyn Riet (University of Toronto)*; Sara Abu-Obaid (Columbia University); Jay Werber (University of Toronto)

Ion-exchange membranes (IEMs) with high conductance and low water permeance are essential for efficient electrodialysis of hypersaline brines; yet, achieving both properties simultaneously remains challenging. Here, we employ thiol-ene photopolymerization to synthesize anion exchange membranes (AEMs) that exhibit exceptionally low water uptake while maintaining high ion exchange capacities (IEC). Film formation included a UV-initiated crosslinking of a reactive pre-polymer, followed by a simple post-modification step to induce fixed cationic charges. The hydrophobic nature of the resulting network combined with high crosslinking density yielded low water uptake and water permeabilities of 0.066 to 0.198 g water/g dry polymer and 9.5 X 10^-14 to 5.4 X 10^-15 m^2 s^-1. This limited water sorption facilitated the formation of AEMs with high charge concentrations ranging from 12 to 25 eq/L. The membranes demonstrated competitive performance metrics, with ionic conductivities up to 13.1 mS/cm and counter-ion/co-ion selectivities exceeding 500 in 1 m NaCl. Through ion transport modelling, we then show that these films, as well as some commercial IEMs, could achieve improved performance if manufactured with thicknesses of 1-2 μm rather than 10-500 μm, which is predominant to date. This work establishes new approaches for fabricating high-performance AEMs, advancing the development of membranes for hypersaline brine treatment.

Pore Structure, High-Throughput Membrane Development, and Compaction-Resistant Membranes for High-Pressure Reverse Osmosis

Sima Zeinali Danalou (University of Toronto); Niher Sarker (University of Toronto); Hongchen Wang (University of Toronto); Patrick Lee (University of Toronto); Jason Hattrick-Simpers (University of Toronto); Jay Werber (University of Toronto)*

High-pressure reverse osmosis (HPRO) extends the operating range of reverse osmosis (RO) from ~80 bar limits typical for seawater RO to 120-150 bar, which would be highly beneficial for brine treatment. However, conventional RO membranes suffer from compaction, in which water permeance goes down by ~50% and pores in the support layer suffer irreversible compression. Over the past three years, our multi-disciplinary team at the University of Toronto has worked with Veolia WTS to understand and mitigate compaction in reverse osmosis membranes at high pressures. This presentation will summarize our work, much of which is being written up for publication.

Key results that we will discuss include:
1. High-Throughput Membrane Development: We are nearly done building a robotic system to automate the synthesis and compression testing of compression-resistant porous membranes. To our knowledge, this will be the first "self-driving lab" in membrane science.
2. Pore structure: We have developed tools to estimate 3-D structures from 2-D SEM images, and developed a method to account for conductive coating thicknesses in SEM, which allows for accurate estimation of real pore sizes and 2-D porosities. Notably, we find that porosities are 2-3-fold higher (25-44% surface porosity) after this adjustment. In other words, the membrane science field has systematically underestimated surface porosity by a large margin.
3. Understanding compaction: We have combined HPRO experiments (up to 150 bar) with pore structure analysis to determine fundamental causes of water permeance loss.
4. Mitigating compaction: Using the above tools and analysis, we have developed membranes with negligible water permeance drop when running up to 150 bar. Further optimization of these membranes is on-going.

Combined, these results will advance the field of HPRO. Our future work aims to translate these findings to other enhanced RO technologies.

Scaleup and Pilot Testing of a High Permeance Membrane for CO2 Capture from a Steel Mill Blast Furnace

David Hopkinson (NETL)*; Lingxiang Zhu (NETL); Thien Tran (NETL); Victor Kusuma (NETL); Grigorios Panagakos (NETL); Glenn Lipscomb (University of Toledo); Neil Pergar (U. S. Steel); Brenda Petrilena (U. S. Steel)

Membrane technologies continue to be of interest for carbon capture applications such as electric power production or steel and cement manufacturing because of their simple operation, modular scalability, and no emissions of amine degradation products. Point source capture from blast furnace stoves in steel manufacturing is an opportunity for membrane technologies because of the higher CO2 partial pressure (0.2-0.3 bar) than other industrial sources, yet higher levels of contaminants may pose a challenge.

Previously we reported the National Energy Technology Laboratory's (NETL's) novel, non-aging rubbery membrane selective material with CO_2 permeability of 930 Barrer with CO_2/N_2 selectivity of 44, which exceeds the 2008 Robeson upper bound. We verified the performance of this membrane material in a 400-hour field test at the National Carbon Capture Center in Alabama, using actual flue gas from a coal-fired and a natural gas-fired boiler. This material exhibited excellent long-term stability in the presence of humidity and contaminants including sulfur particles.

To achieve the highest possible membrane permeance, we used roll-to-roll fabrication equipment to cast thin film composites (TFCs) of our selective material over a custom membrane support. The resulting TFCs were packed in 1000 cm^2 plate-and-frame membrane modules that were designed and 3D-printed in-house. Our lab testing of the modules showed remarkably high CO_2 permeance of >4,000 GPU and CO_2/N_2 selectivity >30.

Beginning in February 2025, we will perform a six-month field test of the membrane modules for CO_2 removal from a steel manufacturing blast furnace at the U. S. Steel Edgar Thomson Plant in Braddock, Pennsylvania; a first-of-its-kind test in the United States. We have constructed a mobile membrane test unit that is being transported to the Edgar Thomson Plant, where the unit can be monitored remotely from NETL. We will present the preliminary field testing results.

Sustainable Polyelectrolyte Nanofiltration Membranes

OLAWUMI SADARE (University of Bath)*; Garyfalia Zoumpouli, (University of Bath); John Chew (University of Bath); Davide Mattia (University of Bath)

Nanofiltration (NF) membranes are traditionally made from fossil-fuel-derived feedstocks and toxic solvents, posing significant environmental and health concerns. This study reports the successful fabrication of stable NF membranes using two biopolymers, cationic lignin, and sodium carboxymethyl cellulose (CMC), and water as the only solvent. The goal of this research is to develop a stable polyelectrolyte membrane (PEM) using cationic lignin and CMC as polycation and polyanion, respectively, via the layer-by-layer (LbL) method, with water as the sole solvent. The PEMs were fabricated on a polyethersulfone (PES) support, and their performance was evaluated at a transmembrane pressure of 2 bar. The membrane's pure water permeance, rejection of charged dyes (methylene blue and reactive orange 16), and molecular weight cut-off (MWCO) were measured. The PEMs exhibited a pure water permeance of 6 LMH/bar at 5 bilayers, with a 93% rejection for methylene blue and 96% for reactive orange 16. The membrane's MWCO ranged from 300 to 620 Da, indicating a loose NF membrane. The mass balance was above 90%, suggesting minimal adsorption. Additionally, the PEMs maintained excellent stability for 30 days in deionized water, attributed to the strong electrostatic interactions between the layers.

This study demonstrates that stable NF membranes can be successfully produced using biopolymeric materials and water as a solvent. The PEM's effective rejection of small, charged molecules makes it promising for applications in protein removal, wastewater treatment, and biotechnology.

Keywords: Polyelectrolyte membranes, nanofiltration, biopolymers, sustainable materials, layer-by-layer.

Membrane/Filter Platforms for Anion Exchange Membranes in Difficult Aqueous Separations

Sam Thompson (University of Kentucky)*; Dibakar Bhattacharyya (University of Kentucky); Jennifer Bukowski (University of Kentucky); Mattie Brock (University of Kentucky)

Performance of separations by ion exchange is influenced by a variety of factors, including charge, concentration, co-contaminants, and reactor design. More specifically, typical ion exchangers suffer performance limitations over time because of poor selectivity, declining efficiencies, and potential formation of channels or plugs, among other concerns. To better understand how the reactor microstructure plays a role in these factors, two alternative approaches for ion exchange separations were analyzed. In this study, porous microfiltration membranes and nonwoven microfiber filters were functionalized with anionic polymers to probe their performance capabilities for the separation of charged small molecules. This functionalization was achieved using two methods of polyelectrolyte deposition: 1) in situ pore functionalization and 2) surface grafting. In situ pore functionalization allows for cationic functionality to be imbued in the membrane via polymerization of amine-containing hydrogel within the pore network. Surface grafting covalently attaches amine residues to the surface of either the pore or microfiber that comprises the material. The mechanism of interaction between functionalized material and passing solute was analyzed for several analytes ranging in hydrodynamic diameters, molecular weights, and electrophilicity/hydrophilicity. This research was supported by NIH funding through the NIEHS-sponsored University of Kentucky Superfund Research Center under award number P42ES007380, and partially supported by Southern Services Company, Atmus Filtration Technologies Inc., and Chevron Co.

Ion Sorption Within Polyphenyl Alkylene Ion Exchange Membranes

Timothy Hudak (The Pennsylvania State University)*; Christopher Gorski (The Pennsylvania State University); Enrique Gomez (The Pennsylvania State Universtiy)

High-performance Ion Exchange Membranes (IEM) are crucial to enabling electrochemical separations such as electrodialysis and membrane capacitive deionization. IEMs must maintain a high degree of rejection of co-ions relative to counterions for the separation to be efficient. Manning's theory of dilute polyelectrolytes has been used to describe co-/counter-ion selectivity in IEMs at thermodynamic equilibrium, with implications for continuum modeling of electrochemical separations; however, success is varied and poor agreement with experiment is often observed in low-water content IEMs. This work aims to: (i) assess the accuracy of the predictions made by Manning's theory for IEMs and (ii) investigate factors contributing to discrepancies observed between experimental results and theory predictions. Using regioselective homopolymer and random copolymer polyphenyl alkylenes, we performed ion sorption experiments for quaternary ammonium functionalized biphenyl, terphenyl, and quarterphenyl homopolymers, and random copolymers. Differential scanning calorimetry was used to quantify the extent of free and bound water for each membrane. Transmission electron microscopy probes membrane morphology, such as the size and distance between hydrophilic domains. Finally, we compare each experimental scenario to predicted partitioning coefficients obtained by Manning's theory, as well as ion partitioning trends as a function of membrane parameters, such as the volumetric fraction of water.

From theory to practice: Challenges in capillary flow porometry for membranes

Dana Dutczak (POROMETER)*; Weronika Motyka (IB-FT GmbH); Loubaba Sharaf Aldeen (IB-FT GmbH); Karolina Chojnacka (IB-FT GmbH); Izabela Struzynska-Piron (POROMETER); Marina Angesleva (POROMETER); Eline Pattyn (POROMETER); Anna Sobolewska (IB-FT GmbH)

Pore size characterization is crucial for evaluating membrane performance and determining its suitability for specific filtration processes. The size and distribution of through pores directly influence the ability of a membrane to separate different substances, affecting both selectivity and permeability. Capillary flow porometry is a widely used characterization technique for measuring pore sizes and pore size distribution in filtration and separation media. It is specifically used to measure the through-pores, which are essential for membrane functionality. However, pore size characterization is associated with many challenges, ranging from sample preparation, through the selection of appropriate wetting liquid and measurement conditions, to the interpretation of measurement curves.

This work addresses these challenges, bridging theoretical measurement principles with practical solutions. It also aims to open discussion on unifying pore size analysis methodology.

Various flat-sheet and hollow fiber membranes were analyzed using gas-liquid (POROLUX™) and liquid-liquid (POROLIQ™) porometers. Special attention was given to sample preparation, wetting procedures, and test parameters such as bubble point determination criteria, flow direction, and measurement runtime. Different sample potting methods and wetting procedures were explored, considering membranes stored in water or glycerol. Methods for removing impregnating agents and liquid exchange techniques were evaluated for their impact on pore size distribution. A novel approach to bubble point determination, independent of the test area, is introduced. Additionally, various measurement curve shapes, their origins, and practical strategies to enhance measurement accuracy are discussed.

These findings provide practical guidance for preparing and accurately measuring different types of membranes with capillary flow porometry, helping to improve pore size analysis and enhance its application in membrane science.

Developing and Assessing the Scale-Up Life Cycle Impacts of Polysulfone Membranes with Chemically Immobilized Silver Nanoparticles and Eco-Friendly Solvents

David Lu (University of Kentucky)*; Isaac Oluk (University of Kentucky); Minwoo Jung (Georgia Institute of Technology); Sophia Tseng (University of Kentucky); Diana Byrne

(University of Kentucky); Tequila Harris (Georgia Institute of Technology); Isabel Escobar (University of Kentucky)

The sustainability of polymeric membrane technology is hindered by the use of hazardous organic solvents. Recent studies on fabricating polymeric membranes using eco-friendly solvents have largely focused on demonstrating proof-of-concept, while the next steps of optimizing performance and investigating scale-up remain understudied.

The first goal of this study was to develop a method to deter microbial growth on polysulfone (PSf) membranes derived from the eco-friendly solvents Rhodiasolv® PolarClean and γ-valerolactone (GVL) by chemically immobilizing silver nanoparticles (AgNPs). The antimicrobial activity of AgNPs is documented, though studies on adding AgNPs to PSf membranes have mainly used physical methods prone to leaching. AgNP-cysteamine nanocomposites were linked to PSf using polydopamine. Characterization analyses validated nanocomposite retention in the membrane. The membrane exhibited similar water permeability to a pristine PSf membrane (21.9 ± 3.1 LMH/bar, 25.0 ± 3.1 LMH/bar, respectively) and minimal leaching.

The second goal was to evaluate the global environmental and health impacts of membrane fabrication using the nanocomposite and a scalable manufacturing technique. Slot die coating (SDC) was selected due to its demonstration of fabricating PSf-PolarClean-GVL membranes on a roll-to-roll system. A life cycle assessment (LCA) was performed to calculate cradle-to-gate impacts of producing flat sheet membranes for 10 impact categories. Membranes made with PolarClean-GVL via SDC generally exhibited higher impacts due to the relationship between solution viscosity and materials inventory. Silver was identified as a major contributing parameter to toxicity and ozone depletion impacts and contributed to greater uncertainty of impacts. Smog, fossil fuel depletion, and acidification impacts were reduced when the electricity source was substituted to a region with a greater percentage of clean energy technologies, highlighting its footprint in impacts.

Identifying key precursor modification parameters for enhancing the separation performance of carbon molecular sieve (CMS) hollow fiber membranes

Brian Leonard (West Virginia University)*

Modular biomass gasification is a promising clean hydrogen (H_2) generation technology and integrating a membrane module with a series of bubbling fluidized bed reactors could enable the separation of high-purity H_2. In this project, a Carbon Molecular Sieve (CMS) hollow fiber membrane created from a Matrimid® precursor is used to effectively separate H_2 from similarly sized molecules such as CO, CO_2, and CH_4. To achieve high-selectivity membranes without requiring extreme processing conditions, we are developing a novel polymer precursor selective layer engineering strategy, and this will be the focus of my talk.

This precursor selective layer engineering strategy will analyze the effects of different diamines on the resulting performance of the CMS membrane. The base conditions will utilize Diethyl toluene diamine (DETDA), which is a mixture of 2 isomers, thus creating different bonding locations for the Matrimid® polymer precursor. The effects of introducing this on the diffusion, sorption and permeation properties of the CMS membrane, will be analyzed and an envisioned mechanism of the CMS microstructure will be presented in my talk. A comparison between DETDA and other diamine entities will be included as well. While the selective layer modification does enhance the selectivity of the CMS membrane, it proved insufficient in achieving the desired purity of the H2 stream (99.98%). Therefore, a synergistic process incorporating changes in final pyrolysis temperature and dwell time has also been investigated and these results will be presented. Finally, the effects of high temperature hydrogen exposure on the sorption, diffusion and permeation properties of the CMS membrane will be briefly discussed.

XPRIZE Water Scarcity: Driving a Desalination Revolution

Hossein Atoufi (XPRIZE Foundation)*

Water scarcity is a growing global crisis, already impacting nearly 25% of the population, with projections indicating a 40% shortfall in water availability by 2030. Coastal regions, where freshwater accounts for just 0.5% of Earth's total supply, face mounting pressure due to rapid urbanization, population growth, and climate change. Desalination offers a viable solution, yet its widespread adoption remains hindered by high costs, inefficiencies, and environmental concerns. Despite being central to modern seawater purification, desalination technology has seen little innovation since the early 2000s. XPRIZE Water Scarcity, a $119 million, 5-year competition, seeks to break this stagnation by driving advancements in desalination systems, materials, and methods to make the process more affordable, energy-efficient, and sustainable. The competition challenges teams to develop novel desalination materials and systems that significantly enhance salt-water separation, cost efficiency, and durability. With oceans holding 97% of Earth's water, harnessing this vast resource efficiently is critical to ensuring global water resilience. XPRIZE Water Scarcity aims to redefine the desalination landscape by fostering groundbreaking innovations that make clean water production more accessible and environmentally responsible. This competition will not only overcome the limitations of existing desalination technologies but also accelerate breakthroughs that transform how the world secures and values water. Through scientific ingenuity, collaboration, and a commitment to sustainability, the competition will unlock a future where clean, potable water is no longer a privilege but a global standard.

Transforming Wastewater Nitrogen into High-Value Fertilizer with Electrodialysis

Mohammed Tahmid (Georgia Institute of Technology)*; Marta Hatzell (Georgia Institute of Technology)

Recovery of nitrogen from wastewater presents a unique opportunity to valorize waste and contribute to a more circular nitrogen economy. However, dilute solution separations are challenging for most state-of-the-art separation technologies. This often results in technologies having low concentration factors that result in low-value products (e.g., < 1 wt % N). Here, we demonstrate how a cascading electrodialysis system combined with a hollow fiber membrane contactor (ED+HFMC) system can achieve efficient recovery of ammonia from simulated centralized animal feeding operation (CAFO) wastewater. We achieved an overall concentration factor of ~200× (~40× in ED and ~5× in HFMC). This resulted in a ~10 wt % NH4+-N fertilizer product. The specific energy consumption (SEC) for the three stages of the ED was

1.89–6.14 kWh/kg NH_4^+-N, which is lower than that of the Haber–Bosch process (8.9–19.3 kWh/kg N). Operating costs were <$0.90/kg NH_4^+-N for each of the electrodialysis stages and NH_3 stripping. Furthermore, the use of renewable sources to generate electricity may reduce GHG emissions by up to 95% compared with the Haber–Bosch process. These results underscore the potential of the ED+HFMC system for ammonia recovery from dilute feed streams, offering both high concentration factors and low energy demand.

Mixed salt sorption in ion exchange membranes: effect of counter-ion identity

Aubrey Quigley (The University of Texas at Austin)*; Jordyn Yoh (The University of Texas at Austin); Nathaniel Lynd (The University of Texas at Austin); Benny Freeman (The University of Texas)

Ion exchange membranes (IEMs) are utilized in various water and energy applications due to their high counter-ion/co-ion selectivity. Although it is well understood that ion transport through dense polymer membranes is largely impacted by ion solubility, few fundamental studies exist that examine equilibrium ion partitioning between a charged membrane and a mixed salt solution. Understanding counter-ion/counter-ion sorption selectivity in IEMs when equilibrated in various salt mixtures can provide insight into the transport properties of charged membranes in complex salt mixtures. Here, systematic ion sorption studies of binary LiCl/KCl and LiCl/$MgCl_2$ salt mixtures were conducted on a commercial cation exchange membrane with fixed sulfonate groups. A range of solution compositions (i.e., mole fractions) were investigated at ionic strengths ranging from 0.01 N to 4 N (i.e., brine conditions), which surpasses the membrane fixed charge concentration. In LiCl/KCl salt mixtures, high selectivity was observed for the more hydrophobic counter-ion (i.e., K^+) due to its heightened affinity to interact with the fixed charge groups. In LiCl/$MgCl_2$ salt mixtures, the divalent counter-ion (Mg^{2+}) possesses an even greater sorption selectivity over Li^+ due to its greater Donnan potential. In both cases, as the external solution concentration is increased and the Donnan exclusion is reduced, the membrane co-ion concentration increases and the counter-ion/counter-ion selectivity decreases. Further, high Donnan exclusion of co-ions prevents sorption of mobile divalent counter-ions (i.e., counter-ions not associated with the membrane fixed charged groups), resulting in a larger fraction of divalent ions associated to the fixed charges and a larger fraction of monovalent ions balancing sorbed co-ions. The findings here provide insight into how solution composition and the extent of Donnan exclusion can impact ion-ion sorption selectivity in mixed salt conditions.

Flow-through permeability apparatus for automated high-throughput analysis of complex solute mixtures

Aubrey Quigley (The University of Texas at Austin)*; Kris Gleason (The University of Texas at Austin); Nathaniel Lynd (The University of Texas at Austin); Benny Freeman (The University of Texas at Austin)

Typically, salt permeability is assessed in diffusion cells, where two chambers, the donor and receiver, are separated by a membrane. The donor chamber has some salt solution of known concentration, while the receiver chamber is filled with pure water. Under an externally applied concentration gradient, ions from the donor chamber permeate across the membrane into the receiver chamber, and the rate of permeate flux can be measured as the receiver chamber conductivity increases with time. While this method is ideal for single salt solutions, it becomes substantially more complex to design, conduct, and analyze these experiments when permeability coefficients of salts in complex mixtures are desired. Since conductivity measurements cannot distinguish between ions of like charge, and in-situ analytical techniques are generally quite specific for particular ions, the most practical way to analyze salt mixture permeability is to take aliquots from the receiver cell over long periods of time to determine the flux of each species. In this work, we have instead designed a novel automated apparatus capable of measuring permeability coefficients of ions in complex salt solutions. This flow-through style cell operates by continuously pumping salt solution through the donor chamber and pure DI water through the receiver chamber. A 6-port isolation valve on the upstream side of the cell allows for multiple mixed salt feed solutions to be analyzed sequentially. On the downstream, permeating ions are continuously collected in the DI water, maintaining a constant permeate composition and concentration. The receiver effluent is directed to an automated fraction collector, which allows the permeate composition to be analyzed as a function of the donor mixture composition. This innovative high-throughput permeability system allows advanced operation of salt transport experiments for more efficient collection of permeability and selectivity data in complex solution conditions.

Transport Modeling in Low-Salt-Rejection Reverse Osmosis (LSRRO) Membranes

Martina del Cerro (Yale University)*; Amit Shocron (Yale University); Hanqing Fan (Rice University); Jack Gilron (Ben-Gurion University of the Negev); Menachem Elimelech (Rice University)

Reverse osmosis (RO) desalination plants are widespread worldwide and are a source of water for stressed regions. However, desalination results in the production of a brine steam that is becoming more challenging to handle due to stringent regulations and high conveyance costs. Moreover, many industries generate wastewaters of high salinity which show promise for reuse but are hard to treat through conventional membrane methods due to pressure limitations. Low-Salt-Rejection Reverse Osmosis (LSRRO) offers an alternative to conventional RO when dealing with hyper-saline solutions. Through leveraging the driving force for permeate flux using Low-Salt-Rejection (LSR) membranes, LSRRO can augment water recovery from brines, while operating at moderate hydraulic pressures.

Description of transport across LSR membranes has traditionally been done using the Spiegler-Kedem-Katchalsky (SKK) model, which incorporates an additional parameter into the transport equations to fully describe water and solute permeation. A drawback in determining SKK transport parameters is that coefficients are concentration dependent. Parameter estimation also relies on accurate depiction of the concentration polarization phenomenon, usually modeled applying the film theory. Accordingly, a mass transfer (MT) coefficient is estimated, even though the assumptions in MT correlations may not hold for transport across LSR membranes.

This work critically reviews conventional approaches for determining transport parameters in LSR membranes and integrates empirical and mathematical optimization methods to obtain SKK coefficients. The developed methodology combines and refines strategies previously employed in parameter estimation, identifying shortcomings of the SKK framework. Through experimentation on commercial LSR membranes, we observe the impacts of pressure on transport and attempt to increase the accuracy of parameter estimation by limiting assumptions and avoiding film theory simplifications.

Development of Rubbery Thin Film Composite Membranes Through Interfacial Crosslinking

Niloofar Shirali (University of Toronto)*; Dean Stipanic (University of Toronto); Jay Werber (university of toronto)

Thin-film composite (TFC) membranes are vital in industrial applications such as gas separation, water purification, and carbon capture due to their high performance and material efficiency. This research explores innovative strategies for fabricating ultra-thin, high-performance selective layers, addressing the challenge of achieving defect-free films with enhanced properties.

Specifically, we expand the chemistries available through interfacial polymerization, focusing on rubbery polymer networks.

The first approach, published in the Journal of Membrane Science (DOI: 10.1016/j.memsci.2024.123532), uses interfacial crosslinking to produce ultra-thin polydimethylsiloxane (PDMS)-based membranes. Amine-terminated PDMS was functionalized with acid chloride groups and crosslinked with polyethylenimine (PEI) to form selective layers, extending traditional polyamide film formation. By optimizing polymer concentrations and reaction conditions, we achieved membranes as thin as 50 nm—>10× thinner than traditional coatings. These membranes showed promising gas permeation properties, with CO_2 permeance of 3290 ± 340 GPU and a CO_2/N_2 selectivity of 12 ± 3, demonstrating their potential for high-efficiency applications.

The second ongoing strategy employs UV-initiated thiol-ene interfacial polymerization to develop ultra-thin rubbery membranes. Thiol-ene photopolymerization offers advantages such as oxygen compatibility, rapid kinetics, and ideal network formation but is underutilized in membrane science. We use poly(1,2-butadiene), a reactive polymer rich in vinyl groups, photo-crosslinked at an oil/water interface with water-soluble dithiols. Preliminary results show rapid formation of defect-free selective layers ~100 nm thick. Future work will explore reaction kinetics and functionalizing these films with tailored chemical properties.

Material Science Advances Towards Functionalized Membranes

Dibakar Bhattacharyya (U. Kentucky)*; S Thompson (University of Kentuckyy); R. Mills (University of Kentucky); F. Leniz (University of Kentucky); M. Leach (University of Kentucky); M. Brock (University of Kentucky); J. Bukowski (University of Kentucky); S. Bosu (University of Kentucky); Noah Meeks (Southern Company)

Functionalization of membranes with responsive polymers and/or nano-catalyst provide applications ranging from water remediation to environmental catalysis to advanced affinity-based remediation to green synthesis. Membranes provide flexible technologies, and various applications range from toxic pollutants removal to biological separations to critical metals. It is also important to address scale-up aspects from lab synthesis commercial viability.

Some questions we plan to answer are: is it possible to make NF type membrane for both critical metals recovery to separation of PFAS type compounds to have lower calcium rejection but very high sugar rejection? What about making NF on Functionalized and responsive MF membrane for both exclusion and capture of trace contaminants? What about making chemical, such as, hydrogen peroxide by non-covalent enzyme immobilization in MF by green chemistry approach. Our technology development included functionalization (green chemistry) of membrane pores with pH and temperature responsive polymers. We address the need for targeted remediation strategies by developing integrated, cost-effective technologies which incorporate both reductive and oxidative strategies in membrane domain without the production of toxic byproducts. This is accomplished through the development of a polymer/MF membrane platform in both lab-scale and full-scale. The presentation will cover: (1) new NF synthesis approaches on functionalized support and applications, (2) lab to module on highly selective sugar type molecule separation with lower divalent metal rejection, (3) direct nanostructured catalyst (enzyme and metals) incorporation in membrane domain and applications. We acknowledge the funding by NSF EPSCoR program, NIH-NIEHS-SRP, and industries.

Nanoengineering Polyorganosilica (POSi) Porous Structure via Few-cycle Atomic Layer Deposition for Superior H2/CO2 Separation

Vinh Bui (University at Buffalo)*; Amandine Tirino (University at Buffalo); Ameya Tandel (University at Buffalo); Erda Deng (University at Buffalo); Won Il Lee (Stony Brook University); Leiqing Hu (University at Buffalo); Lingxiang Zhu (National Energy Technology Laboratory); Chang-Yong Nam (Brookhaven National Laboratory); Haiqing Lin (University at Buffalo)

Polyorganosilica (POSi) membranes with excellent H_2 separation properties can be rapidly prepared via 2-minute oxygen plasma treatment of polydimethylsiloxane (PDMS)-based thin-film composite (TFC) membranes. However, POSi membranes are subject to hydrothermal instability, decreasing H_2 permeance and H_2/CO_2 selectivity. In this study, we employ a few-cycle atomic layer deposition (ALD) of aluminum oxide to induce metal-siloxane covalent linkage, which can simultaneously stabilize POSi networks and tighten porous structures to improve H_2/CO_2 separation performance. First, oxygen plasma treatment induces an ultrathin (<10 nm) POSi top layer with abundant silanol (Si-OH) groups, as confirmed by XPS and SEM images. Second, the hydroxide groups enable the ALD process, where exposure to trimethylaluminum (TMA) forms Al-O-Si covalent linkages. The effect of the ALD parameters on the POSi network structure, gas transport properties at 150oC, and hydrothermal stability are thoroughly examined. For example, exposing POSi membranes to 3-cycle ALD improves H_2/CO_2 selectivity from 67 to 380 but decreases H_2 permeance from 380 to 230 GPU. When challenged with roughly 0.6 mol% water vapor, the pristine POSi membrane exhibits an irreversible drop of H_2/CO_2 selectivity to 55. At the same time, a 3-cycle of ALD retains the selectivity, suggesting improved hydrothermal stability with the ALD treatment. The superior H_2/CO_2 separation properties and robust hydrothermal stability coupled with the facile fabrication processes demonstrate the potential of the POSi membranes for practical H_2 purification and CO_2 capture.

Desalination of High Salinity Brines Using Single-Stage Osmotically Assisted Reverse Osmosis

Glogina Boakye (Auburn University)*

Desalination plants and other industries produce hypersaline brines and frequently dispose these through non-ecofriendly methods. Surface water discharge, the most common and least expensive of these discharge methods, negatively impacts aquatic life at the point of discharge through osmotic imbalance and thermal stress. Although several advanced methods have been introduced to treat hypersaline brines towards zero/minimal liquid discharge, high capital and operational costs challenge implementation. While reverse osmosis (RO) is theoretically cost-effective for treatment of hypersaline brine, conventional commercial polymeric RO membranes cannot handle the hydraulic pressure necessary to overcome the osmotic pressure of hypersaline brines. To observe any permeation in a RO setup, a minimum pressure of ~70 bar must be applied while the burst pressure of current RO membranes is ~80 bar. In this work, we propose a Single-Stage Osmotically Assisted Reverse Osmosis (SSOARO) method, a modified Osmotically Assisted Reverse Osmosis (OARO) process, to treat hypersaline brines. This is aimed at treating hypersaline solutions in a single stage to minimize the energy and capital demand while improving the pure water recovery. The strategy of this setup is to permanently encapsulate a draw solute within two stacked RO membranes to reduce the osmotic pressure of the system. The SSOARO system rejected 64% NaCl at a flux of 1.9 Lm-2h-1 using a hypersaline feed of 1.54 M NaCl and a hydraulic pressure of 70 bar. Under the same conditions, a salt rejection of 72% and flux of 2.1 Lm-2h-1 was observed for a conventional RO membrane. The lack of improvement in flux using our SSOARO system has led us to probe the transport mechanisms. Accordingly, our work delves into the effects of both diffusive and advective transport modes. Each transport mode tends to dominate at different points within the membrane and at different conditions of pressure, temperature, and concentration.

Zwitterionic Membranes for Rare Earth Element (REE) Separations

Nafiseh Khoshnevisan (Tufts University)*; Ayse Asatekin (Tufts University); Ashleigh Herrera (Tufts University); Christine Duval (Case Western Reserve University); Venkat Ganesan (University of Texas at Austin)

Rare-earth elements (REEs) play a critical role in renewable energy technologies, such as wind turbines, as well as in electric vehicles, green energy generation, electronics, and high-performance airframes due to their electrochemical, magnetic, alloy-strengthening, and luminescent properties. Currently, REEs are extracted from mined ore and purified through hundreds of sequential solvent extraction stage which is costly, and environmentally harmful, generating toxic waste and consuming large amounts of energy. Replacing these high-energy, complex technologies with membranes could dramatically decrease both costs and environmental impact associated with REE refining. However, currently available membranes cannot effectively separate REEs from each other. In this study, we aimed to design membranes with functional nanopores formed by the self-assembly of copolymers containing zwitterionic (ZI) ligands to enable selective separation of REEs. Thin-film composite (TFC) membranes with cross-linkable zwitterionic amphiphilic copolymers (X-ZACs) as selective layers were prepared by coating a thin copolymer layer onto a porous support. The effective pore size, quantified by the filtration of neutral solutes, can be tuned by adjusting the degree of crosslinking by post-processing steps that polymerize or connect double bond side-groups on the X-ZACs. Our preliminary data indicates that these membranes can have effective pore sizes that range from ~1.5 nm down to <1 nm, with the smallest pore sizes resulting in ion selectivity. Membrane selectivity for adjacent rare-earth elements can be controlled by tuning the pore chemistry and pore size using X-ZAC membranes. Our preliminary data indicates that X-ZAC membranes with specific ZI monomers can show differences in the rejection of different REEs and other cations, opening the door to future applications.

Evaluation of NH3 Diffusivity and Solubility in ZnCl2 Immobilized Molten Salt (IMS) Membranes

Moses Adejumo (Clarkson University)*; Abdulkadir Zabi (Clarkson University); Simona Liguori (Clarkson University)

Ammonia (NH3) production is widely recognized for its role in fertilizer manufacturing and emerging use as hydrogen carrier. Conventionally, NH3 is produced via the Haber-Bosch (H-B) process at 400 – 450 °C and 25 – 30 bar. Due to the low conversion associated with this process, the NH3 produced has to be separated from the unreacted N2 and H2 via condensation and refrigeration, with a negative impact on the carbon footprint. Switching the traditional NH3 separation technology to alternative methods like membrane separation can help mitigate CO2 emissions and reduce energy consumption. Among classes of ammonia-permeate membranes, inorganic membranes offer superior thermal stability and promising separation performances at high temperatures. ZnCl2 immobilized molten salt (IMS) membranes have shown selectivity of NH3 over N2 and H2 up to 107 and moderate permeance of ~200 GPU at 300 °C and atmospheric pressure. To better understand the NH3 permeation through the molten salt, this study experimentally determines the solubility and diffusion coefficients of NH3 in the ZnCl2 IMS membrane at ~300 °C and atmospheric pressure. The solubility coefficients were determined by conducting sorption experiments using a modified thermogravimetric analyzer, while a time-lag method using on-line mass spectrometry was adopted to determine the diffusion coefficient of NH3 in the mixture across the IMS membrane. The results revealed that NH3 solubility decreases with increasing NH3 partial pressure in the feed, exhibiting carrier saturation phenomenon. The diffusion coefficient from time-lag increases with increasing NH3 partial pressure in the feed. The data generated from sorption and time-lag experiments were used to determine permeability and compared with the one calculated directly from flux permeation measurements.

Beyond the First Shell: A Free Energy Outlook on Dehydration-Controlled Ion Transport in Polyamide Membranes

Hengyu Xu (University of Wisconsin-Madison)*; Subhamoy Mahajan (University of Wisconsin-Madison); Ying Li (University of Wisconsin-Madison)

Membrane-based ion separation is essential to modern society, playing critical roles in the water, energy, and medical industries due to its energy efficiency and versatility. Although polymer membranes have been widely employed for ion separation, the fundamental mechanisms of ion transport behavior remain unrevealed, which hinders theoretical breakthrough of the permeability-selectivity trade-off in polymer membranes. Here, the ion transport phenomena within polyamide membranes are investigated from an energy perspective through umbrella sampling (US) simulations. The simulation results indicate that both monovalent and divalent ions form hydrated structures with 2 hydration shells in solution. Inside the polyamide membrane, partial dehydrations of both the 1st and 2nd hydration shells occur. While, the dehydrations of the 1st and 2nd hydration shells are dominated by the binding and steric effects, respectively. Therefore, beyond the size and charge of hydrated ions, the density of polar functional groups and dynamic pore structure of polymer membrane also influence ion transport inside the membrane through binding and steric effects. Additionally, results from 2D US simulation show a flatter energy-hydration number curve within the polymer membrane, implying an easier dehydration or rehydration inside the membrane. Consequently, ion transport resistance at the membrane surface is attributed to the directional dehydration induced by steric effect, whereas, resistance within the membrane arises from the distribution of polar functional groups. On this basis, a model is proposed to evaluate the Gibbs free energy of ion transport by disentangling the contributions of binding and steric effects. The rejection predicted by this model shows excellent agreement with experimental results. This work provides molecular insights into ion transport across polymer membranes, facilitating the design of functional polyamide membranes with enhanced ion selectivity.

Salty no More: Explainable AI-Driven Discovery of Polyamide Desalination Membranes

Subhamoy Mahajan (University of Wisconsin-Madison); Ying Li (University of Wisconsin-Madison)*

As water scarcity tightens and demands surge, the urgency of ultra-selective seawater desalination grows. Reverse-osmosis using polymeric membranes such as polyamide offers energy-efficient solutions with rejections as high as 98-99%. However, significantly improving the water permeability while maintaining the salt rejections remains an open challenge. Since the debut of polyamide membranes, the monomers (trimesoyl chloride and m-phenylenediamine) have remained unchanged due to their unparalleled performance. Some advances in identifying polymer chemistry have been performed, albeit most studies mainly focus on improving fouling and chlorine resistance by changing the surface chemistry which is an important operating constraint. In this work, we focus on exploring the impact of polyamide chemistries on desalination performance. An extensive dataset of water permeability and salt rejection is generated using molecular dynamics simulations of ~1000 distinct amine chemistries (trimesoyl chloride is kept unchanged) with ~150 μs of non-equilibrium pressure gradient simulations. The dataset is used to assess the performance of a wide array of machine learning (ML) models such as artificial neural networks, graph neural networks, and graph convolution neural networks. Features such as Morgan fingerprint with frequency, Alvadesc molecular descriptors, graph representation of amine monomers, and porous media properties from molecular simulations are used. The importance of chemical and structural information is determined using SHAP (Shapley Additive exPlanations) analysis, to make the ML explainable. Finally, the trained ML model is then used to explore a 7 million database of amine monomers to identify membranes with high water permeation and salt rejection, which is verified using additional molecular dynamics simulations.

What is the Role of Relative Humidity on Conductivity in Polymer Electrolytes?

Nico Marioni (The University of Texas at Austin)*; Akhila Rajesh (The University of Texas at Austin); Rahul Sujanani (University of California, Santa Barbara); Zidan Zhang (The University of Texas at Austin); Leo Gordon (University of California, Santa Barbara); Raphaële Clément (University of California, Santa Barbara); Rachel Segalman (University of California, Santa Barbara); Benny Freeman (The University of Texas at Austin); Venkat Ganesan (The University of Texas at Austin)

Ion transport properties are widely studied in polymer electrolytes under rigorously dry and highly swollen conditions. While our knowledge of ion transport in the extreme cases of dry and highly swollen polymer membranes is relatively well developed, the transition between these extremes, i.e., the low hydration regime, is poorly understood. Recently, our experimental collaborators published a method to measure ionic conductivity in salt-doped polyethers under controlled humidity conditions (less than 10% water by volume). Interestingly, the conductivity exhibited two distinct regimes of transport, slow and fast, as a function of water content, which appear to be correlated to the lithium-water solvation characteristics. However, a detailed understanding of the mechanistic origins of this result is unknown. In this study, we apply atomistic molecular dynamics simulations to probe the role of hydration on ion conduction in LiTFSI-doped polyethers under 0-80% relative humidity conditions (0-10% water by volume). Our results suggest that lithium conduction within this limit can be divided into slow and fast regimes. Under dry conditions, lithium conduction is dictated by the relative motions of polymer ether oxygen solvation cages. Under humid conditions, lithium ions are preferentially solvated by water molecules. At low water contents, the average lithium is weakly hydrated (two or less water molecules), maintaining strong interactions with its solvation cage and therefore exhibiting a negligible increase in conduction relative to dry conditions. At higher water contents, lithium ions form complete hydration shells, residing within water-rich domains that promote fast conduction, significantly increasing the overall conductivity. By contrast, TFSI has comparatively weak interactions with its environment, exhibiting only a marginal increase in conduction at all hydrations probed due to the increase in local dynamics.

Affinity membranes containing complexing agents for Li+/Mg2+ separation

Jada Mowatt (University at Buffalo, The State University of New York)*; Sagnik Das (University at Buffalo, The State University of New York); Haiqing Lin (University at Buffalo, The State University of New York)

Lithium (Li) is a highly desired element for clean technology, and its production from salt lake brine often involves the separation of lithium chloride (LiCl) from its mixtures with magnesium chloride (MgCl2)). Nanofiltration (NF) membranes present an exciting alternative method to extract lithium, being cost-effective, energy-efficient, and environmentally friendly. However, there is a tradeoff between Li+ permeance and Li+/Mg2+ separation factor (SF); high permeance is typically faced with a low separation factor and vice versa. Herein we demonstrate that NF membranes can be surface-modified with chelating agents, such as ethylenediaminetetraacetic acid (EDTA), to achieve superior Li+/Mg2+ selectivity, by forming strong complexes with divalent Mg^{2+}, as opposed to monovalent Li^{+}. Specifically, commercial NF270 and NF90 were first primed using bio-adhesive dopamine, followed by EDTA grafting via its reaction with polydopamine (PDA). The membrane surfaces were characterized using scanning electron microscopy (SEM), water contact angle (WCA), and atomic force microscopy (AFM). The effect of the dopamine coating time, EDTA concentration, and feed solution pH on Li^{+}/Mg^{2+} separation properties was thoroughly investigated. The EDTA grafting increases the Li^{+}/Mg^{2+} SF. For example, the modification using PDA2h-EDTA3h slightly increases the Mg^{2+} rejection from 99.1% to 99.9% but dramatically increases the Li^{+}/Mg^{2+} SF from 68 to 1260 at pH 3. Additionally, the WCA decreased from 60.3° (NF270) to 42.8° (PDA2h-EDTA3h), indicating increased hydrophilicity. Both single salt and mixed salt tests were completed and showed consistent trends. The performance of NF90 was also compared with that of NF270. Overall, this study unveils the potential of harnessing the affinity with ions to improve the membrane separation properties.

Green Chemistry with Coupled Electrocatalysis and Electrodialysis: enhancing Selectivity and Conversion Efficiency

Kaylin Hicks (University of Arkansas)*; Jamie Hestekin (University of Arkansas)

Carbon neutrality is an important goal for many industries, particularly in biopharma and commodities. Achieving this goal requires a multilayered approach that includes recycling and upcycling of materials, material replacement, and material minimization. A key strategy is to source materials sustainably by replacing fossil fuels with bio-based alternatives and utilizing CO_2-neutral processes. In this project, we focus on producing organic acid monomers, which are important for producing polymers. Currently, production relies on fermentation converting bio-based sources with genetically engineered bacterial strains or fossil fuel sources. This research takes an approach that challenges the disadvantages of fermentation by developing a flow-based process that catalytically converts a biomass source to organic acid products and combines catalysis with selective separation for enhanced efficiency and product recovery. The project goal is to recover lactic acid (LA), the organic acid of choice, selectively in high yield (>70%) from glucose conversion via coupled electrocatalysis and electrodialysis membrane separation. Experimentation was conducted using a multistep process. LA separation is controlled from a mixed media stream, based on previous results/data from LA conversion by electrocatalysis, with activated wafer electrodionization (AWEDI) membranes. Lastly, a batch system was created and tested, coupling the two processes, utilizing Cu foil as direct points to power the Cu/ion exchange resin wafer for conversion in the feed chamber of the AWEDI. Different ratios of Cu to ion resin wafers (50%, 60%, and 80% Cu) were made to test the increase/decrease of conversion and separation. Currently, in the combined system, LA yield is around 5-10% using the 50% Cu wafer. This work shows that a green chemistry coupled system driven by electrochemistry can work to produce LA. The yield can be increased with further experimentation of wafer chemistry and process design.

Microporous polymer membranes for ion and organic molecules separations

Sui Zhang; Wong Kar Chun (National University of Singapore)*

Microporous polymers, such as conjugated microporous polymers (CMPs) and covalent organic networks (COFs), present a versatile platform with rich pore geometry, functionality for ion and molecular separations.

In this presentation, I will present our efforts in preparing CMP and COF membranes, and exploring their applications for shape-based separation of molecules and Li+/Mg2+ separations. Through controlling the monomers and diffusivity in the media, CMP membranes with different pore geometry and distribution were prepared, enabling shape-based selectivity. In addition, by matching the pore size and wall functionality of the COF channels, an interaction-accelerated mixed ion separation was achieved, reaching a Li/Mg mixed ion selectivity over 1000.

Concentration of organic solvents by osmotically assisted reverse osmosis

Mengyang Hu (Kobe University)*; Hideto Matsuyama (Kobe University)

Organic solvents are essential in many chemical and pharmaceutical manufacturing processes. However, the expansion of chemical production has led to the generation of substantial quantities of organic solvent wastewater. The presence of these organic solvents in wastewater has exacerbated environmental pollution. Current treatment practice involves direct combustion and distillation. Besides leading to waste generation and high energy consumption, the processes do not promote a circular economy. This issue can be effectively mitigated by implementing organic solvents recycling processes. Here a unique, osmotically assisted reverse osmosis (OARO) strategy was utilized to effectively concentrate and treat organic solvent wastewater for recycling of the solvents. The influence of several operational parameters, such as flow rate, applied pressure, number of membrane modules, and their connection mode, was systematically investigated. Specifically, enhancing the operating pressure and reducing the feed flow rate will both result in a higher concentration of organic solvents. Furthermore, we developed a simulation model that can reproduce and predict the OARO process. Investigating long-term operational stability revealed that the OARO process can sustain stable operations for isopropanol (IPA) and N,N-dimethylformamide (DMF) for approximately 6 months and 18 months, respectively. Finally, real organic solvent wastewater was treated using the OARO technique, resulting in a successful concentration of IPA and DMF by approximately 22 and 42 times, respectively. The findings of this research underscore the significant practical applicability of the OARO strategy for organic solvent wastewater recovery, recycling, and reuse.

A Versatile Polymer Precursor for Carbon Molecular Sieve Gas Separation Membranes

Ching-En Ku (University of Maryland); Gaurav Iyer (University of Maryland); Yutong Wu (University of Maryland); Chen Zhang (University of Maryland)*

Carbon molecular sieve (CMS) membranes are made by pyrolysis of polymer precursors and have shown highly competitive separation performance for many gas separation applications. Traditionally, different polymer precursors are required to fabricate CMS membranes for target separations. For example, precursors with low fractional free volume (e.g., polybenzimidazole and cellulose) can provide CMS membranes with attractive H2/CO2 selectivity, whereas precursors with higher fractional free volume (e.g., fluorinated polyimides) are needed to fabricate CMS membranes for larger gas pairs such as olefin/paraffin. It is highly desirable to identify a polymer precursor that can give CMS membranes attractive for a wider spectrum of gas separations by simple change of pyrolysis conditions. Such a versatile precursor would allow the fabrication of CMS membranes for diverse gas separations using the same precursor, thereby streamlining CMS membrane development and production process.

In this talk, we will present a versatile copolymer precursor that can give CMS membranes with highly attractive H2/CO2, CO2/CH4, and olefin/paraffin separation performance by simple change of pyrolysis temperature. High-temperature (over 1000 °C) pyrolysis of the copolymer precursor gives CMS membranes with ultra-high H2/CO2 selectivity over 700 under aggressive high-temperature mixture permeation, which significantly exceeds the polymer upper bound. Low-temperature (550 °C) pyrolysis of the copolymer precursor gives CMS membranes with outstanding CO2/CH4 and olefin/paraffin separation performance exceeding the polymer upper bounds. An investigation of CMS membrane pore structure suggested that its copolymer backbone was responsible for its unique versatility as a CMS membrane precursor.

Parametric Optimization of Layer-by-Layer Coated Ion Exchange Membranes: Insights from Computational Modeling

Omar Elshawarby (KU Leuven)*; Xing Yang (KU Leuven)

The authors chose not to publish the abstract.

Investigation of Hydrogen Permeation and Material Characterization of Pd-Ag-Y Membrane

Omid Jazani (Clarkson University)*; Simona Liguori (Clarkson University)

There is an urgent need for clean fuels due to global warming. Hydrogen is considered a clean fuel for many applications. Pd based membrane reactors could be used to produce ultra-pure hydrogen with zero or low carbon emission, depending on feedstock. The aim of this study is to investigate the hydrogen permeation properties of the novel Pd82-Ag15-Y3 membrane in the presence of different supports and gas mixtures. The tests were performed at 400 °C and 1.4 – 3 bar. The unsupported membrane showed one of the highest hydrogen permeability in literature. Results showed that the unsupported membrane had a range of 12-267% higher hydrogen permeation than supported membranes. Thus, the presence of porous supports used to enhance the mechanical stability of the membrane negatively affected the hydrogen permeation due to mass transfer limitation. Binary hydrogen permeation tests were conducted in the presence of N2, CH4, CO2, and CO with different concentrations at 400 °C by using unsupported and supported membranes. The results revealed that CO had the highest inhibition effect for all the unsupported and supported membranes tested due to competitive adsorption on the surface. No superficial adsorption on the membrane was observed for N2, CH4, and CO2 during permeation at 400 °C. The membrane characterization was performed by scanning Electron Microscope (SEM), Electron Backscatter Diffraction (EBSD), and X-ray diffraction (XRD) on both pristine and used membranes. The results showed that the Pd82-Ag15-Y3 membrane is a good candidate to be used in small-scale hydrogen production and separation.

Synthesis and Water Vapor Separation Properties of Azine-linked COF Membranes

Jose Lopez-Cazares (Arizona State University)*; Jerry Lin (Arizona State University); Kailong Jin (Arizona State University)

Two-dimensional covalent organic frameworks (2D COFs) are attractive candidates for advanced molecular separation due to their robust, uniform, and tunable porous structures. This work reports on the fabrication of microporous ACOF1 membranes by substrate-assisted liquid-liquid interfacial polymerization in synergy with solvent-assisted vapor annealing to study the condensable vapor/gas permeation properties. Separation of water vapor and air mixture is studied to understand their condensable vapor transport properties and how their separation performance correlates with the microstructure of the ACOF1 selective layer in terms of level of crystallinity and crystallite orientation. The solvent-assisted annealed ACOF1 membrane composed of randomly oriented crystallites offers high water vapor/oxygen separation because of water capillary condensation which is triggered at relatively low humidity range (~20—25 %) at room temperature due to their improved crystallinity and appropriate intrinsic ~1-nm ACOF1 pores and advantageous azine linkage-water interactions. The annealed ACOF1 membranes had water vapor permeance ranging from 6 x10^-8 to 2.2x10^-7 mol/s m^2 Pa, and separation factor as high as 710. This research work can help to incentivize studies of microporous 2D COF membranes applied to membrane-based dehumidification and water harvesting, which have been scarcely explored.

Techno-economic assessment of a carbon molecular sieve-based xylene isomer purification process

Hammed Balogun (Georgia Institute of Technology)*; Conrad Roos (Georgia Institute of Technology); Ryan Lively (Georgia Institute of Technology)

Xylene isomer purification is an important separation in the chemistry industry as it is critical to the production of polyethylene terephthalate (PET). Commercially available technologies for this separation are complex and are often energy-intensive. Recently, membrane materials capable of separating this isomeric mixture without a phase change have been developed and potentially provide a low-energy footprint technology with simpler operation. While there has been no large-scale application of membrane-based xylene isomer separation, this study conducts a techno-economic analysis (TEA) to evaluate the commercial attractiveness of a p-xylene selective carbon molecular sieve (CMS) membrane. First, single-stage organic solvent reverse osmosis (OSRO) and pervaporation (PV) processes were used to evaluate the influence of membrane properties and process parameters on separation performance. At low p-xylene recovery, pervaporation system outmatches OSRO with a reduction of ~23 kJ energy required per kg of p-xylene product. However, PV module heating compared to constant feed pressurization in OSRO resulted in a 5 – 6 times energy requirement at high recovery for PV systems. Finally, two membrane cascade systems–a single PV stage followed by two OSRO stages and a three-stage OSRO, were benchmarked against the commercial Parex process. Results show a cost savings of 52- 85% for the optimal PV/OSRO cascade system compared to the commercial Parex process, demonstrating the commercial attractiveness of membrane systems in small molecule solvent separations.

Polymer-Silica Caged Biocatalytic Porous Media for Biofouling Control in Anaerobic Membrane Bioreactors

Syed Salman Ali Shah (Beijing Normal University at Zhuhai)*

Anaerobic membrane bioreactors (AnMBRs) are increasingly favored for their low energy requirements, but membrane fouling is a major barrier to their broader application (Drews, 2010; Shah et al., 2022). Quorum quenching (QQ) using live microbial cells immobilized in hydrophilic polymer matrices (e.g., hydrogels) has been effective for mitigating biofouling, but challenges in mechanical durability and long-term stability remain. This study addresses these challenges by developing a composite biomedium that encapsulates QQ bacteria (Rhodococcus sp. BH4) within porous silica cages. The silica cages protect the cells from solvent toxicity, and polysulfone reinforcement enhances structural integrity, providing thermal and chemical resilience.

Surface characterization confirmed successful encapsulation of the QQ bacteria within the silica framework. The composite's unique architecture integrates viable QQ bacteria, amorphous silica, and polymer networks, ensuring flexibility and sustained bacterial activity. The reinforced biomedia exhibited significantly improved mechanical properties, with tensile strength nearly four times that of traditional hydrogel-based systems, offering durability for extended use.

In solvent exposure tests, unprotected *Rhodococcus sp. BH4* cells were severely inactivated, while silica-encapsulated cells preserved 70% viability and 77% QQ activity, highlighting the protective role of silica. In anaerobic membrane bioreactor trials, the composite media achieved superior biofouling control, extending operational durations approximately three-fold compared to reactors without media and two-fold compared to reactors with empty media. The composite also reduced biopolymer content and quorum-sensing signal concentrations, extending membrane lifespan without compromising treatment efficiency. This work advances antifouling QQ media design, enabling long-term retention of functional live cells in robust polymer-silica hybrid systems.

Mechanistic Understanding and Modeling of AAV Full and Empty Capsid Separation using Multi-modal Ligands

Xiaolei Hao (University of Arkansas); Shu-Ting Chen (University of Arkansas); Ranil Wickramasinghe (University of Arkansas); Xianghong Qian (University of Arkansas)*

Viral vectors such as adeno-associated virus (AAV) particles are efficient gene delivery carriers. Unlike protein-based therapeutics, the development of purification platforms for viral vectors is far less ma¬ture. For example, the separation of full and empty AAV capsids has been particularly challenging due to their similarities in size and physiochemical properties. Previously we have fabricated multi-modal anion exchange (MMAEX) membranes for separating the empty AAV capsids from the full capsids with high resolution. Here the effects of mobile phase properties such as ionic strength, pH and buffer type as well as the stationary properties such as the chain density and chain length of the ligands on the empty/full capsid interaction with the ligand were investigated. The mechanisms for their binding interactions are elucidated and critical parameters identified. Moreover, a mathematical model was developed to describe full and empty capsid chromatographic separations. This theoretical framework could help guide future chromatographic viral particle separations.

Organic solvent reverse osmosis membrane for cyclic siloxane purification

Hammed Balogun (Georgia Institute of Technology)*; Reza Haghpanah (The Dow Chemical Company); Dimitris Katsouli (The Dow Chemical Company); Jay Rose (The Dow Chemical Company); William Koros (Georgia Institute of Technology); Ryan Lively (Georgia Institute of Technology)

Siloxanes and silicone polymers such as silicone fluids, elastomers, and resins are important chemicals with a wide range of applications in the biomedical, electronic, cosmetic, and space industries, etc. Conventional techniques for the separation of the small siloxane molecules from silicone fluid such as stripping and adsorption are complex and energy intensive. Membrane separations could serve as alternatives to or complement these industrial separation techniques to reduce the overall process energy demand. This work presents the application of polymeric membranes in the purification of siloxane oligomers via the rejection of silicone oil contaminants. First, the fundamental transport properties are studied to understand the driving force and mechanism for the separation. These properties are then incorporated into a Maxwell-Stefan (MS) framework for siloxane permeation and separation predictions. Experimental analysis shows that the diffusivities and solubilities of the siloxanes decrease with increasing Si atoms, and the resulting siloxane and oil fluxes are accurately matched with the solution-diffusion (SD) model using these independently-measured parameters. Unit activity analysis confirms that sorption selectivity dominates the SD separation performance with SD4/SOil = 1250 and DD4/DOil = 122. Finally, an octamethylcyclotetrasiloxane (D4) cleansing operation using polydimethylsiloxane (PDMS) membranes was carried out with a feed containing 90/10 wt/wt% D4/silicone oil. With a transmembrane pressure of 20 bar, a high-purity D4 permeate stream (>99wt% D4) was obtained. The discrepancy in SD flux prediction was found to be due to concentration polarization with a concentration polarization modulus of 4.2. Modification to the MS framework could improve solvent/polymer mixture transport prediction in polymeric membranes. These findings provide an industrially scalable low-cost alternative to silicone fluid separation and siloxane purification.

The influence of spacer arm identity on salt transport in charged nanofiltration membranes

Laurianne Lair (University of Notre Dame)*; Jonathan Ouimet (University of California, Santa Barbara); William Phillip (University of Notre Dame)

There is a growing demand for materials that can separate solutes of similar size and charge, such as critical minerals. Current processes including liquid-liquid extraction (LLE) apply molecular recognition principles which leverage ligand-solute coordination to facilitate separations. Unfortunately, LLE requires copious amounts of harmful solvents to achieve the desired separation. Self-assembled copolymer membranes provide a framework for post-synthetic modifications to incorporate molecular recognition ligands. The versatility of this platform presents a promising avenue for identifying design rules for membranes capable of producing high purity products with reduced negative environmental impacts. This study investigates the influence of ligand identity on ion transport in copolymer membranes. Using a copolymer membrane scaffold, ligands with varying chemistries are covalently bound to the pore wall through a copper(I) catalyzed azide-alkyne reaction. The ligands contain a carboxylic acid group attached to the membrane by spacer arms with varying numbers of alkyl (i.e., 4, 6) or ether (i.e., 0, 2, 4, 6, 10) repeat units. Due to the presence of ionizable carboxylic acid groups, the solution pH and ionic strength influence membrane performance. High-throughput experiments are conducted to systematically analyze the effect of feed composition and membrane nanostructure on the transport of two model salts, NaCl and Na2SO4. For instance, these salt transport experiments highlight that ligand identity does not affect NaCl transport. However, the transport of Na2SO4 increases with the length of the ether spacer arm while decreasing with the length of the alkyl spacer arm. The result scales with the relative hydrophilicity, determined through water contact angle measurements and water volume fraction calculations, of the membrane. Overall, the spacer arm length and chemistry are additional important degrees of freedom to consider when designing ion-selective materials.

Organosoluble Pentiptycene-containing Polybenzoxazole Membranes for High Temperature Gas Separations

Sandra Weber (University of Notre Dame)*; Agboola Suleiman (University of Notre Dame); Ruilan Guo (University of Notre Dame)

The development of high-performance, processable polymer membranes for high-temperature gas separations is needed to integrate separations into reaction processes, such as H2 separation after olefin dehydrogenation, to improve process economics and energy efficiency. Conventional polymer membranes often lose their size-sieving capability at elevated temperatures due to increased segmental chain motion. Crosslinked polymers or polymers with rigid backbones, like thermally rearranged (TR) polybenzoxazoles (PBOs), show promise, but their lack of solution processability limits industrial use. Herein we present a one-pot, silylation-assisted, in-situ cyclodehydration method to produce linear, organo-soluble pentiptycene-containing PBOs (LsPPBO) that can be solution-cast into thin films. In contrast to insoluble TR-PPBOs that show a ~72% decrease in H2/CH4 selectivity from 35°C to 150°C, LsPPBO films demonstrate only a ~36% selectivity reduction over the same temperature range. Moreover, the ~128% increase in permeability with temperature displayed by LsPPBO moves the performance laterally approaching the 2008 H2/CH4 upper bound. Additionally, doping the LsPPBO film using phosphoric acid enables over a 30% improvement in size-sieving, surpassing the 2008 upper bound under mixed-gas conditions. To further enhance high-temperature tolerance, a strategy of semi-interpenetrating networks (sIPNs) was employed, wherein high molecular weight LsPPBO chains thread the rigid model networks formed by end-crosslinking of LsPPBO oligomers to simultaneously provide stability and maximize permeability-selectivity combinations. This sIPN design allows for systematic tuning of crosslink density and network composition and corresponding performance. Mixed-gas H2/CH4 permeation tests show sIPNs with high crosslink density outperform the 2008 upper bound across a wide temperature range, demonstrating the potential of solution-processable PBO membranes for high-temperature gas separations.

Jammed Interconnected Bilayer Emulsions (JIBEs): A New Biomimetic Material for Ion-Selective Membranes

Aida Fica (University of Texas)*

Biomimetic membranes hold great potential for selective separations, yet most research has focused on aquaporin-based desalination, leaving ion-selective membranes largely unexplored. This is due to the challenge of fabricating defect-free membranes, which is critical for achieving high selectivity in ion transport.

Here, we introduce Jammed Interconnected Bilayer Emulsions (JIBEs) as a novel biomimetic material for ion-selective membranes in electrically driven processes. JIBEs consist of billions of polydisperse aqueous compartments stabilized by amphiphilic monolayers that jam together to form a robust network of bilayers, mimicking the architecture of biological tissues. This gel-like material exhibits mechanical and electrical properties akin to biological membranes, supports tunable ionic transport through incorporated ion channels, and uniquely features self-healing, allowing the bilayer network to reversibly reform after disruption.

As a proof of concept, we developed a JIBE-based membrane functionalized with gramicidin, a peptide ion channel selective for monovalent cations ($NH_4^+ > K^+ > Na^+$). This macroscopic membrane preserved the transport properties of nanochannels, achieving K^+/Na^+ selectivity of 5.5 and NH_4^+/Na^+ selectivity of 16.8, comparable to intrinsic gramicidin selectivity. In contrast, control membranes without gramicidin exhibited negligible transport, confirming the formation of a defect-free network where ion transport occurs exclusively through embedded channels.

This demonstration highlights JIBE membranes' scalability, ease of protein incorporation, stability in water, and self-healing nature, paving the way for highly selective biomimetic membranes. Future developments incorporating highly selective ion channels could enable biomimetic membranes with transport properties dictated solely by embedded channels, revolutionizing ion separations.

Fluorocarbon Reduction via Semifluorinated Copolymer Thin Films by Combining Spin Coating and Ring-Opening Metathesis Polymerization

Matthew Vasuta (Vanderbilt University)*; Zane Parkerson (Vanderbilt University); Tyler Oddo (Vanderbilt University); Bridget Rogers (Vanderbilt University); Kane Jennings (Vanderbilt University)

Amorphous perfluoropolymer thin film composite membranes drastically reduce the energy requirement over thermal methods for separating water from a polar organic, but their syntheses pose significant health concerns from their excessive use of per- and polyfluoroalkyl substances (PFAS). We report a method to substantially reduce the amount of PFAS, solvent, and time needed to synthesize a low surface energy polymer membrane through the copolymerization of norbornene (NB) with 5-(perfluoro-n-alkyl)norbornenes (NBFn) in a single process that combines spin coating with ring-opening metathesis polymerization (scROMP). The scROMP approach efficiently integrates polymer film synthesis and deposition into one rapid process, converting monomer into polymer films in < 2 min with < 1 mL of solvent for a 36 cm^2 film. Perfluoroalkyl chain lengths, n, of 4, 6, and 8 were examined, with the fluorocarbon component tending to dominate the surface for all n, exhibiting water contact angles comparable to the fluorocarbon homopolymer even with as little as 2% NBFn in the contacting monomer. These semifluorinated copolymer films were used in ethanol dehydration as low PFAS substitutes for amorphous fluoropolymer membranes. Even 7% fluorocarbon in the polymer (or 2% in the monomer) produced an order-of-magnitude increase in selectivity over a fully hydrocarbon membrane, with additional fluorination up to 63% (50% in monomer) leading to another order-of-magnitude enhancement and properties similar to the pNBFn homopolymer. Additionally, the dense outer fluorocarbon layer provides an ideal setup to estimate the sorption and diffusion components of selectivity for fluorocarbon and hydrocarbon groups within a membrane.

Atomistic Insights into the Reactive Diffusion of CO_2 in TMG-Based Facilitated Transport Membranes

Changlong Zou (The Ohio State University)*; Yang Han (The Ohio State University); Li-Chiang Lin (National Taiwan University)

In pursuing advanced CO2 separation technologies, facilitated transport membranes (FTMs) with engineered chemistry have emerged as a leading solution for their capability to enhance both permeance and selectivity. These membranes utilize mobile carriers, such as 1,1,3,3-tetramethylguanidine (TMG), to react with CO2 and form carbamate and bicarbonate, each following distinct diffusion pathways. The inherent complexity of the reactive diffusion of CO2, coupled with the need for molecular-level resolution in understanding the diffusion behavior, necessitates employing sophisticated computational tools.

In this study, we employ advanced molecular dynamics (MD) simulations with a re-parameterized reactive force field (ReaxFF) to investigate the reactive diffusion of CO2 in TMG under water vapor conditions. By analyzing the trajectories of CO_2, TMG, and their complexes, we identify two primary diffusion mechanisms: coupled diffusion and hopping diffusion. In coupled diffusion, bicarbonate ions formed from the CO_2-TMG reaction move in tandem with TMG molecules, resulting in a diffusion coefficient similar to that of TMG. In contrast, hopping diffusion involves a rapid transfer of bicarbonate ions between adjacent TMG molecules, notably increasing the bicarbonate mobility compared to systems lacking this hopping process. This increased mobility further facilitates the CO_2 transport across the membrane.

These findings improve our understanding of CO_2 transport in FTMs with mobile carriers, emphasizing the roles of coupled and hopping diffusion in enhancing membrane performance. This computational approach offers a cost-effective alternative to extensive experimental methods and aids the design of efficient, selective membrane materials for CO_2 separation.

Separation and Recovery of Rare Earth Elements from EV magnets using Membrane Solvent Extraction System

Shailesh Singh Dangwal (Oak Ridge National Lab)*; Zachary Coin (Oak Ridge National Lab); Syed Islam (Oak Ridge National Lab); Ramesh Bhave (Oak Ridge National Lab)

Rare earth elements (REEs) particularly dysprosium (Dy), praseodymium (Pr), and neodymium (Nd) have drawn an urgent attention worldwide and have become strategically critical for the security of global economy, energy, transportation, and communications. Majority of REEs are imported and are not domestically produced in United States. This puts the entire supply chain of REEs at a large geopolitical risk. On the other hand, widespread applications of REEs in different devices such as portable electronic and especially hybrid/electric vehicles (EV) generate millions of tons of end-of-life (EOL) waste containing metallic contaminants. Recycling of REEs from EOL EV magnets can help in mitigating the supply side risk of REEs and can also address the environmental challenges arising due to EV waste containing metallic contaminants. In this work, we report the separation and recovery of high purity REEs form multiple EoL EV magnets with varying composition using scalable, energy-efficient, and environmentally friendly membrane solvent extraction (MSX) system. In MSX, extractant is immobilized inside the pores of hollow fibers and it interacts selectively with the target elements by making a complex and thus facilitates the separation of the target elements. In the first stage of MSX, REEs (Dy, Nd, and Pr) are separated from non-REEs (Iron, Copper, Zinc, and Boron etc.) with >99.5 wt.% purity and >95% recovery using Diglycolamide as an extractant. Thereafter, in second stage, heavy REE (Dy) is separated from light REEs (Nd and Pr) using phosphonic ion based extractant with >99.9 wt.% purity and >95% recovery. The purity of the recovered product was further confirmed with additional characterization such as SEM, EDS, and XRD. Overall, MSX technology provides a potential pathway for industrial deployment to domestically produce of REEs in United States.

Designing Ion-Exchange Membranes for Vanadium Redox Flow Batteries

Gregory Reimonn (University of Michigan Kamcev Lab)*; Po-Han Huang (University of Michigan Kamcev Lab); Zackary Mitzel (University of Michigan Kamcev Lab); Jovan Kamcev (University of Michigan Kamcev Lab)

Vanadium Redox Flow batteries offer a promising solution for renewable energy storage but face efficiency and capacity challenges due to poorly understood transport processes in ion-exchange membranes. These membranes must conduct charge balancing species efficiently while blocking vanadium crossover, which compromises performance. Addressing this challenge requires fundamental understanding of membrane properties, internal solution equilibria, and ionic transport phenomena. In this study, we synthesized ion-exchange membranes with systematically varying membrane properties and employed in-situ Raman spectroscopy for acid-base and ion-pairing equilibria within membranes equilibrated in relevant electrolytes. Combining transport and equilibrium characterization revealed that proton exchange reactions between acid species within the membrane promote conductivity in excess of ideal transport models, even in membranes with significant residual hydration. Ion-paired species formation limits conductivity by closing conduction pathways between acid species. Key outcomes of this work include new models for proton conduction in acidic environments, rational design parameters for future membrane development, and novel methods to characterize ion pairing in ion-exchange membranes. These findings advance both theoretical understanding and practical approaches for optimizing Vanadium Redox Flow batteries.

Biocatalytic Membrane for the Degradation of Diisononyl Phthalate

Hadi Rouhi (University of Alabama)*; River Copeland (The University of Alabama); Ryan Summers (The University of Alabama); Steven Weinman (The University of Alabama)

Diisononyl phthalate (DINP) is a commonly utilized plasticizer in plastic manufacturing. Over time, DINP can leach from plastic products into the environment, accumulating in dust, air, and water. The U.S. Environmental Protection Agency finalized a regulation designating DINP as a toxic chemical, highlighting its environmental and health concerns. Enzymes, as protein-based catalysts, facilitate specific biochemical reactions, making them valuable in various applications. Lipase from Candida rugosa (LCR) is notable for its role in biotransformation, including the biodegradation of phthalic acid ester plasticizers. Enzyme immobilization offers a strategy to create continuous membrane bioreactors. In this study, LCR was covalently immobilized onto a polyvinylidene fluoride (PVDF) membrane. The membrane was functionalized through a redox-initiated graft-from polymerization process using an amine-based monomer, introducing primary amine groups to the surface. These amine groups were reacted with glutaraldehyde, facilitating the subsequent LCR attachment. Functionalization in each step was supported by FTIR, water contact angle, and BET surface area analysis. The functionalized biocatalytic membrane was employed to degrade DINP. The membrane's degradation efficiency exhibited a strong positive correlation with enzyme concentration, increasing from ~1% to over 92% as LCR concentration rose from 0 g/L to 10 g/L. Electrospray ionization (ESI) analysis identified monoisononyl phthalate (MINP), phthalic acid (PA), and isononyl alcohol (INA) as primary degradation products. Kinetic analysis using the Michaelis-Menten model yielded a Km of 53.27 and a Vmax of 38.41, providing insights into the enzymatic degradation process. This study demonstrates the potential of immobilized LCR on PVDF membranes for the effective degradation of DINP, offering a sustainable approach to addressing environmental contamination by phthalates.

The unusual transport of phenolic compounds in biorefinery streams through polymer membranes

Woo Jin Jang (Georgia Institute of Technology)*; Ryan Lively (Georgia Institute of Technology)

Bio-oils produced by biomass upgrading processes, including fast pyrolysis, contain valuable chemicals that can be processed into value-added chemical feedstocks and fuels. However, separating and refining these oils remains challenging, as conventional thermal methods often induce polymerization or oxidation of the bio-oil compounds. Membrane-based separations can potentially address this problem by avoiding high temperature operation. Unfortunately, most membrane materials and processes struggle to recover dilute compounds from complex mixtures as well as separate solutes from other solutes. Here, we use a crossflow permeation system with a novel spirocyclic polytriazole membrane (DUCKY-9) to separate a synthetic bio-oil mixture of guaiacol, water, methanol, acetol, and glucose via pressure-driven liquid phase permeation. Our initial hypothesis was that the larger guaiacol and glucose compounds would be partially rejected by the DUCKY-9 thin film composite (TFC) membrane based on the molecular weight cut-off of the DUCKY-9 membrane. Remarkably, the membrane selectively permeates guaiacol—even though it is highly dilute and the second-largest molecule. We find that guaiacol's flux is strongly coupled with methanol's flux, enabling "uphill" permeation of guaiacol in terms of membrane driving forces without second law violations. We hypothesize that this phenomenon stems from solubility and chemical affinity between the polymer and solvents, and we demonstrate that tuning this flux coupling can dramatically enhance separation performance. Notably, increasing the feed's methanol content by only 5 mol% boosts guaiacol's selectivity over glucose from 10 to 25, despite their similar sizes (i.e. 124 Da vs 180 Da). Our experiments suggest that this uphill membrane permeation process may enable recovery of diluted and valuable compounds from complex mixtures.

Interpretation of membrane transport experiments via the solution-diffusion model in the limit of low pressure

Kevin Reimund (University of Texas at Austin)*; Benny Freeman (University of Texas at Austin)

The solution-diffusion (SD) model provides a simple and unifying methodology for describing penetrant transport through non-porous polymeric membranes. The model predicts that penetrant flux across a membrane is proportional to the penetrant concentration gradient across the membrane and that flux is generally indifferent to how this gradient is maintained, e.g., via hydraulic pressure or via osmotic pressure. However, when applied to reverse osmosis, the SD model is commonly represented as $Jw = A(\Delta P - \Delta\pi)$ for water flux, and $Js = B\Delta c$ for solute flux, where ΔP and $\Delta\pi$ are the transmembrane hydraulic and osmotic pressure differences, and A and B are ostensibly intrinsic membrane properties. In this simplified form, flux is proportional to ΔP, rather than to the transmembrane concentration gradient. In recent years, a number of studies have suggested that the SD model provides an inaccurate description of membrane transport and have proposed that the assumptions underlying the model are fundamentally flawed.

In contrast, in this work, we examine the analytical behavior of the SD model applied to reverse osmosis in the limit of low driving force, i.e., as ΔP goes to 0. We show that the interpretation of A as the hydraulic permeance requires potentially unrealistic assumptions about the sorption behavior of the membrane. We show that in the limit of low pressure, even when water flux is linearly proportional to ΔP, the sorption behavior cannot necessarily be neglected. Specifically, in the low-pressure limit, flux is not proportional to A, but rather A scaled by a correction term related to the water sorption isotherm which may only be neglected under certain conditions.

Consequently, the material properties derived from the simplified model may be inaccurate when the isotherm correction term is erroneously neglected. In these cases, the simplified SD model will be unable to accurately predict membrane performance under different operating conditions.

Anionic covalent organic framework membranes for selective PFAS removal and dye desalination

Thanh-Tin Nguyen (University of Wisconsin-Milwaukee); Rahul Sampat Khandge (University of Wisconsin-Milwaukee); Xiaopeng Min (University of Wisconsin-Milwaukee); Peng Hu (University of Wisconsin-Milwaukee); Ho-Kuan Yu (University of Wisconsin-Milwaukee); Hanlun Gao (University of Wisconsin-Milwaukee); Yin Wang (University of Wisconsin-Milwaukee); Xiaoli Ma (University of Wisconsin-Milwaukee)*

Covalent organic frameworks (COFs) are promising materials for nanofiltration due to their tunable pore structures and surface functionalities. This presentation discusses two new applications of ionic COF membranes for selective nanofiltration. In the first study, we demonstrate the use of continuous anionic TpPa-SO_3H COF membranes for the selective removal of per- and polyfluoroalkyl substances (PFAS), addressing the issue of excessive salt rejection associated with conventional nanofiltration membranes. Thin TpPa-SO_3H COF selective membrane layers were fabricated on polymer supports using a scalable counter-diffusion interfacial polymerization (IP) method to create thin film composite (TFC) membranes. These membranes achieved effective PFAS rejection (99% for PFOS; 90–95% for PFOA and short-chain PFAS) while allowing the passage of scale-forming salts, enabling high salt/PFAS selectivity. The membranes also demonstrated superior water permeance (19.9–37.5 LMH/bar), surpassing commercial benchmarks like NF270. In the second study, we constructed mesopores in anionic COF membranes and examined them for dye-salt separation (or dye desalination), a challenge in textile wastewater management. The COF membranes were initially synthesized at a liquid-liquid interface and then transferred to a polymer support to form a TFC membrane. The mesopores, approximately 2.8 nm in size, along with a strong negative membrane surface charge, resulted in a salt-dye selectivity greater than 400 across a wide pH range (3–11.5) under crossflow conditions. This high selectivity facilitates the simultaneous removal of dyes and recovery of salts, addressing both pollution control and resource reuse challenges.

Leveraging dielectric exclusion to improve the desalination transport properties of hydrated polymers by engineering water/polymer interactions

Sean Bannon (University of Virginia)*; Rachel Fetter (University of Virginia); Geoffrey Geise (University of Virginia)

Many desalination processes use polymer-based membranes to control rates of water and salt transport. Critical to controlling these transport rates is to engineer interactions between water molecules and the polymer. One strategy to engineer these interactions is to distribute hydrophilic functionalities through the polymer matrix to restrict water molecule dynamics and to reduce the molecular polarizability of water in the polymer relative to water in the bulk. This phenomenon reduces the dielectric constant of the polymer, which contributes to excluding salt from partitioning into the polymer. While this so-called dielectric exclusion phenomenon is favorable for desalination, it remains largely misunderstood because there are few studies that characterize both polymer relative permittivity (i.e., dielectric constant) and transport properties, and classic models that describe this phenomenon have not captured appropriately the relationship between polymer dielectric constant and transport properties.

Recently, we developed an updated Born model-based approach to describe dielectric exclusion in hydrated polymers and to show how polymer morphology, as well as the dielectric constant, influence dielectric exclusion. To test our model, we synthesized linear and crosslinked co-polymers containing systematically-varied concentrations of hydrophilic hydroxyl functionalities to control polymer morphology (i.e., mesh size) and water molecule dynamics in the polymers. We characterized their relative permittivity and transport properties and found that crosslinking the polymers and increasing the concentration of hydroxyl functionalities in the polymer matrix increased the extent of dielectric exclusion as described by the updated Born model. This presentation presents these results, which guide molecular engineering strategies for polymeric desalination materials.

Selective Ion Transport in Precision Polyanion Membranes

Kyoungmin Kim (Florida State University)*; Chinnadurai Satheeshkumar (Florida State University); Sebastian Castro (Florida State University); Joseph Schlenoff (Florida State University); Justin Kennemur (Florida State University); Daniel Hallinan (Florida State University)

Concentrated brine, the by-product of desalination, can threaten the ecosystem by disrupting the osmotic balance of marine species. The toxicity of heated brine from thermal-based desalination facilities is more severe. Membrane-based brine mining using ion-exchange membranes (IEMs) offer a promising solution for safe water treatment with significant economic benefits and minimized environmental impact. Key challenges lie in the selective ion separation performance, particularly the trade-off between ion selectivity and permeability, and high water-sensitivity of the ionic conductivity. Higher permeability allows for greater throughput, but it could result in the transport of undesired ions which may require additional purification. Also, membranes with increased fixed charge density which provides enhanced hydration and ionic conductivity can dissolve in water. Insoluble membranes can be produced through crosslinking or the use of block copolymers, however, this increases the volume of the inert domain, which limits hydration and reduces ionic conductivity. To develop applicable membranes with sufficiently high ionic conductivity, it is essential to increase the polymer's hydrophobicity while simultaneously enhancing the dynamics of functional groups and ensuring the connectivity of conductive domains. We developed a sulfonated precision polymer membrane. The unique architecture of the polystyrene-like polymer with five repeating carbons instead of two in the backbone, enabled flexible high strength IEMs. The water solubility of the polymer was controlled by carefully tuning the degree of polymer ionization. The IEMs achieved a high ionic conductivity of over 20 mS/cm in the presence of LiCl. Using different counterions resulted in interesting behaviors in water uptake and ion transport. The distinct transport behaviors associated with different forms of IEMs will shed light on the application of precision polymer IEMs in selective ion separation process.

Validation of high-flux plate-and-frame membrane module design for CO_2 capture: a CFD Approach

Cheick Dosso (Carnegie Mellon University)*; Hector Pedrozo (Carnegie Mellon University); Thien Tran (National Energy Technology Laboratory); Lingxiang Zhu (National Energy Technology Laboratory); Victor Kusuma (National Energy Technology Laboratory); David

Hopkinson (National Energy Technology Laboratory); Lorenz Biegler (Carnegie Mellon University); Grigorios Panagakos (Carnegie Mellon University)

This work focuses on developing advanced plate-and-frame membrane modules for CO_2 separation from industrial point sources using CO_2-selective flat sheet membranes. We use Computational Fluid Dynamics (CFD) modeling as we aim to achieve uniform flow distribution, minimize dead-end zones, and mitigate concentration polarization effects, which often limit the efficiency of gas separation membranes. Our goal is to improve membrane module designs by integrating material properties with operational conditions to enhance performance.

Our methodology integrates Multiphysics modeling, combining fluid flow dynamics with the transport of concentrated species within membrane systems. Initial studies have shown excellent agreement between experimental data and simulated results, particularly regarding CO_2 recovery, gas composition in permeate and retentate streams, and stage-cut performance.

Building on this foundation, we have successfully scaled up our design and increased membrane surface areas to accommodate larger flow rates and improve CO2 separation. While earlier designs featured a single 24 cm^2 membrane operating at flow rates up to 600 sccm, the scaled-up modules now support up to 5 membranes with a total surface area of 120 cm^2 and flow rates as high as 3200 sccm. Experimental validation of these enhanced modules confirms the reliability of our approach.

Further achieved advancements include extending our models to a more complex gas feed composition of CO_2, N_2, and O_2. The results are very promising, as shown by negligible relative errors between experimental and simulated data – the maximum relative error on the CO2 recovery was calculated to be 3%. By combining CFD insights with experimental data, this study provides a pathway to optimizing CO_2 separation processes, focusing on both module design and overall process efficiency.

Nodular networks in hydrated polyamide desalination membranes enhance water transport

Yuexiao Shen (Texas Tech University)*; Danyang Li (Tsinghua University); Menachem Elimelech (Rice University); Xia Huang (Tsinghua University)

Thin-film composite reverse osmosis membranes have been the gold standard technology for desalinating seawater or reusing wastewater for nearly half a century. Surprisingly, despite the importance of these membranes to address global water scarcity worldwide, the actual structure of the polyamide selective layer of the membranes under hydrated conditions and its relationship to water and salt transport remain poorly understood. Employing cryo-electron tomography, we successfully reconstructed the three-dimensional surface structure of six commercial polyamide membranes under hydrated conditions for the first time, revealing a fully swollen nodular network. The highly heterogeneous nodules, measuring 17.2±2.8 nm in thickness, were directly connected to the pores of the underlying polysulfone substrate. The nodules occupied most of the surface area compared to the 75.9±26.8 nm thick dense layer of the polyamide film, in contrast to our traditional understanding of these membranes. Key structural parameters of the nodules, including surface area index, wall thickness, and polymer density, were well-correlated with the water permeance of additional 16 polyamide desalination membranes, validating the major role of these nodules in water transport. This study enhances our understanding of desalination membranes and how their heterogeneous structure governs membrane transport properties.

Integrating CFD and Optimization Under Uncertainty in High-Flux Membrane Design for Carbon Capture

Cheick Dosso (Carnegie Mellon University)*; Hector Pedrozo (Carnegie Mellon University); Lorenz Biegler (Carnegie Mellon University); Grigorios Panagakos (Carnegie Mellon University)

Carbon dioxide (CO_2) emissions from industrial sources are a significant driver of global greenhouse gas emissions. To mitigate these emissions and achieve net-zero targets by 2050, carbon capture technologies have become increasingly important. In particular, membrane-based separation has emerged as a promising alternative due to its energy efficiency and scalability.

This study presents a superstructure optimization framework for membrane-based carbon capture. We build reduced-order models using data from simulations of relevant membrane modules. These reduced models incorporate key operating variables, including inlet flow rate, retentate and permeate pressures, and inlet CO_2 concentration, while predicting CO_2 recovery and permeate purity. In this way, we facilitate the selection of the most effective multi-stage membrane process configuration. It should be noted that this optimal design problem includes both the optimal configuration of the multi-stage membrane process and the optimal operating conditions of the process.

Beyond superstructure optimization, we also apply a methodology to account for variability in inlet CO_2 and N_2 concentrations and flow rates during early design steps. In this regard, we develop a two-stage stochastic programming formulation to address this challenge, where the objective is to minimize the expected capture cost while accounting for the probabilistic distribution of key parameters. This approach provides a more realistic assessment of process performance and cost by incorporating variations in feed conditions beforehand. The developed framework contributes to the systematic design of robust and cost-efficient carbon capture strategies that can adapt to a range of operating environments, ensuring the target capture rate and product purity across different operating regimes of a point source.

Architecting High Performance H2/CO2 Separation Membranes: Synergistic Polymer Blends of m-PBI with Microporous Polymers

Joseph Emery (Notre Dame)*; Ruilan Guo (Notre Dame)

H2/CO2 separation using polymer membranes for pre-combustion CO2 capture requires materials with robust thermal stability and high separation performance at elevated temperatures. While polybenzimidazoles (PBIs), notably m-PBI (Celazole®), exhibit excellent size-sieving properties. Their application is hindered by low H2 permeability due to dense polymer chain packing. Limited success has been achieved in developing new PBIs with enhanced H2 permeability without compromising selectivity, particularly at elevated temperatures. Polymer blending offers a convenient yet powerful strategy that allows for the synergistic combination of high selectivity from m-PBI with high permeability from select microporous polymers, resulting in enhanced overall performance. Herein we report the development of two series of m-PBI based blend membranes with attractive H2/CO2 separation performance, which incorporate either a novel pentiptycene-based PBI (PPBI) or a triptycene-containing partially alicyclic polyimide (Trip-AATC), identified through active machine learning (ML) modeling. m-PBI/PPBI blends leverage shape-persistent configurational free volume of pentiptycene units within PPBI creating interconnected microcavities that facilitate H2 diffusion without significantly disrupting the m-PBI's size-sieving capabilities. As a result, a 50% m-PBI/PPBI blend is ~3.5 times more permeable surpassing the 180ºC H2/CO2 upper bound under mixed-gas conditions. The integration of m-PBI with Trip-AATC, an ML-inspired alicyclic polyimide exhibiting exceptional H2/CO2 separation performance, introduces intriguing dual-phase morphologies, leading to all m-PBI/Trip-AATC blends approaching the 180ºC H2/CO2 Robeson upper bound under mixed-gas conditions. This talk will highlight the fundamental relationships between polymer blend morphology and gas transport properties in these m-PBI-based blend membranes, demonstrating the power of targeted polymer blending for advanced gas separation membranes.

Polyimide Hollow Fiber Membrane Contactors for Air Humidification in Fuel Cell Systems: Optimization of Membrane Morphology and Transport Properties via Wet-Gap Spinning

Liang-Chih Ma (Parker Hannifin)*; Stephen Telford (Parker Hannifin); Simon Padron (Parker Hannifin)

Membrane humidifiers present a promising solution for air humidification in proton exchange membrane fuel cell (PEMFC) systems. Asymmetric polyimide (PI) hollow fiber membranes are ideal for membrane humidifiers due to their high water vapor permeability, low air permeability, and superior chemical, thermal, and mechanical stability. However, controlling the membrane morphology and structure to achieve optimal performance through nonsolvent induced phase separation (NIPS) process remains challenging. The wet-gap spinning technique offers an effective approach to modifying membrane outer surface morphology and structure by governing the membrane formation pathway. Gaining deeper insights into the selection of external fluid for polyamic acid (PAA) precursor fiber spinning is essential for enhancing process efficiency and achieving the desired morphology and structure of PI hollow fiber membranes used in gas-gas membrane contactors for air humidification. This study investigates the effects of external fluid solubility, flow rate, air gap distance, and take-up speed on membrane morphology and gas transport properties. The PI hollow fiber membrane contactors were tested at 60°C using a feed stream of 90% RH wet air and a sweep stream of dry air, with controlled retentate and permeate pressures. Water vapor permeance was determined through Wilson plot analysis and N_2 permeance was measured via single-gas permeation at room temperature. The results indicate that the applicable flow rate is limited by the solubility of external fluid, while the membrane's outer surface pore size and porosity are strongly influenced by the external fluid flow rate. By optimizing spinning parameters with a suitable external fluid, a defect-free PI hollow fiber membrane with an ultrathin selective layer and a highly porous bulk structure was successfully developed. The membrane exhibits an excellent water vapor permeance exceeding 5,000 GPU while maintaining N_2 permeance below 0.2 GPU.

Modeling Analysis of Membrane Reactor for Methane Steam Reforming for Hydrogen Production - A Comparative Study on CO2- and H2-Selective Membranes

Oscar Ovalle-Encinia (Universidad Autónoma Metropolitana - Iztapalapa); Jerry Y.S. Lin (ASU)*; ParthKumar Patel (ASU)

Hydrogen is a crucial feedstock for the chemical industry and is also used to power vehicles and generate electricity. Steam methane reforming (SMR) is a well-established and efficient method for hydrogen production. In recent years, membrane reactors have gained significant attention for enhancing SMR efficiency. Hydrogen-permselective Pd-Ag membranes have shown great potential for pure hydrogen production, while more recent developments in CO_2-permeable membranes offer an alternative approach for achieving high-purity hydrogen separation.

This paper presents a comparative analysis of membrane reactors utilizing H_2-selective Pd-Ag membranes and CO_2-selective ceramic-carbonate dual-phase membranes for SMR-based hydrogen production. To ensure a fair comparison, a simulated study is conducted using a consistent reactor model. Specifically, the study models natural gas steam reforming in a tubular Pd-Ag membrane reactor for CO-free hydrogen production and compares it with a previously developed CO_2-permeable membrane model under identical operating conditions.

Simulation results illustrate the effects of key parameters, including the permeation number (θ), which represents membrane H_2 permeance, and the Damköhler number (Da), which characterizes the steam reforming reaction rate. Additionally, the impact of feed pressure on both the reaction and sweep sides is analyzed to assess reactor performance. A parametric study compares the two membranes in terms of H_2 yield, CH_4 conversion, retentate CO concentration, and permeation ratio. Results indicate that both membrane reactors exhibit similar performance at higher feed pressures. At a feed pressure of 5 atm and a vacuum sweep of 10^{-3} atm, with a permeation number of 3 and a Damköhler number of 10,000, both reactors achieve an H_2 yield exceeding 99.7% and a permeation ratio above 99.9%.

Novel Organic Solvent Nanofiltration Membranes for Hydrocarbon Separations

Meltem Urgun-Demirtas (Argonne National Laboratory)*; Lauren valentino (Argonne National Laboratory); Chaoyi Ba (Argonne National Laboratory)

Organic solvent nanofiltration (OSN) membrane separations present a lower-energy, scalable, and intensified alternative to conventional thermal separation processes, which in turn reduces costs and greenhouse gas emissions. OSN technology has been used in various industrial productions such as recovering active pharmaceutical ingredients, solvents from crystalline mother liquors, natural product isolation. This paper focuses on simultaneous membrane material and process development for OSN technology for hydrocarbon separations. Argonne has recently developed a non-fluorinated PIF-based (PIF-X) platform technology, which has multiple key merits such as low-cost, cross-linkable, functionalizable, and highly thermally/chemically stable in harsh environments for various membrane applications. A thin layer of PIF-X polymer was coated onto a PAN support membrane. Commercially available membranes (Borsig and Evonik membranes) were used to establish a baseline for key performance metrics: chemical stability, permeability, and selectivity. Six commercially available OSN membranes and new OSN membranes were characterized through FTIR and contact angle measurements. FTIR analysis allows for the identification of functional groups present in the membranes and provides a greater understanding of the membrane composition. This is useful to anticipate any potential interactions with organic solvents. Contact angle measurements provide a quantitative assessment of the surface hydrophilicity/hydrophobicity, which is related to solvent affinity and permeation characteristics. OSN membrane performance was investigated for separating a nine-component hydrocarbon blend under various operating pressures.

The performance of PIF-based OSN membranes exceeded commercial membranes (10-20% increase in selectivity and 20-40% increase in permeance). The PAN-PY UF support membrane costs $72/m2, much less expensive (3-5x) than the commercial OSN membranes.

Green(er) Solvents for Upcycling Waste PVC into Membranes

ATTA UR RAZZAQ (The University of Alabama)*; Milad Esfahani (The University of Alabama)

The increasing levels of plastic pollution have led to a growing threat to wildlife and ecosystems. One step towards sustainability is upcycling plastic waste into high-value-added products, such as membranes for water treatment. However, the use of organic solvents in the upcycling process of waste plastic into membranes raises serious concerns due to their adverse environmental and health effects. To enhance the eco-friendliness and sustainability of this process, this study investigates the use of green(er) solvents, such as dihydrolevoglucosenone (Cyrene) and triethyl phosphate (TEP), for upcycling waste PVC pipes into ultrafiltration membranes. The rheological and thermodynamic behavior of waste PVC pipe casting solutions with these two solvents were analyzed, along with the properties (pore morphology, size, surface charge, and surface roughness) of the upcycled PVC membranes. Additionally, the effect of additives such as calcium carbonate and carbonyl group possessing compounds present in the waste PVC pipe on the casting solution, membrane structure, and, consequently, on the performance of the upcycled membranes was assessed. The upcycled PVC membranes were benchmarked against similar membranes fabricated using organic solvents and commercial membranes. The additives affected the pore morphology and surface charge of the upcycled membranes. The upcycled PVC membranes exhibited permeability (26.1 LMH/bar with Cyrene and 56.9 LMH/bar with TEP) and dye rejection (94.6 % with Cyrene and 72.6 % with TEP) comparable to similar membranes fabricated using organic solvents and commercial membranes.

Upcycled PVC Membranes from Waste Plastics: Evaluating the Potential for Additive Leaching (Dioctyl Phthalate)

ATTA UR RAZZAQ (The University of Alabama)*; Milad Esfahani (The University of Alabama)

Upcycling waste plastics into high-value membranes is a promising approach that enhances the sustainability of membrane fabrication while helping to mitigate plastic pollution by converting waste plastics into functional products. However, as many plastics contain plasticizers and additives from their initial production, the stability of these additives in upcycled membranes requires careful investigation due to potential health concerns. During the manufacturing of flexible PVC (F-PVC) pipes, various additives such as dioctyl phthalate (DOP), a potentially harmful chemical, and calcium carbonate are incorporated to enhance their flexibility, durability, strength, and workability. A key concern in the upcycling process is the stability and leaching behavior of these additives during the process and later from the upcycled product (membrane), as this directly impacts the feasibility and safety of the entire upcycling process. In this study, we conducted a comprehensive investigation into the transport and leaching behavior of dioctyl phthalate (DOP) and calcium carbonate during the upcycling of waste flexible PVC (F-PVC) into ultrafiltration (UF) membranes via the phase inversion process. Furthermore, we evaluated the stability and leaching of these additives from the upcycled PVC membranes under filtration conditions, assessing both dynamic (1-month) and static (7-month) scenarios. The results revealed that DOP did not leach out during the upcycling process of membrane synthesis and that the dynamic and static leaching of DOP from the F-PVC membranes was negligible, measuring less than 0.006 ppm. This concentration level was well below the limits set by the World Health Organization and the United States Environmental Protection Agency for DOP in drinking water. Additionally, the water permeability and dye rejection of the membranes remained consistent throughout the 1-month duration under a pressure of 100 psi, indicating stable membrane performance.

Strategies to mitigate antifoam-induced membrane fouling. Effect of membrane chemistry and operating conditions on flux and transmission decline.

Federico Volpin (Novo Nordisk)*; Bianca Schroth (Novo Nordisk)

Novo Nordisk is a global pharmaceutical company with over 50 years of expertise in biomanufacturing. Today, it produces half of the world's insulin and a number of other biopharmaceutical products.

In biomanufacturing of protein/peptides, antifoaming agents are frequently used due to the rapid agitation and intense aeration required in fermenters.

There are a wide range of antifoams in the market, from silicone oil to vegetable/mineral oil in the form of cationic, anionic and non-ionic surfactants.

Most antifoams interact with the surface of membranes either reversibly or irreversibly. However, cationic antifoams and silicon oils in particular are known to have negative impacts to mem-brane performance and cleanability. This is both due to the hydrophobic interactions between the membrane surface and antifoam hydrophobic tail, and/or due to pore blocking when the antifoam is in micelle form. As the membrane surface becomes more hydrophobic, it ceases to provide permeate flux, and the process comes to a stand-still.

In this presentation, we share the results of our investigation on the effects of antifoams on different membrane chemistries (PES, regenerated cellulose, PVDF, ceramic ZrO2/TiO2) and how operating parameters like temperature, crossflow, and TMP can affect the product mass flux and fractionation efficiency. Membrane cleanability and the effects of different CIP strategies are also discussed.

Pretreatment to Enhance the Performace of Virus Filtration

Ranil Wickramasinghe (University of Arkansas)*; Xianghong Qian (University of Arkansas); Hironobu Shirataki (Asahi Kasei Medical); Andrew Zydney (Pennsylvania State University)

Virus filtration is used to ensure a high level of virus clearance required in the manufacture of biopharmaceutical products such as monoclonal antibodies. Flux decline during virus filtration can occur due to the formation of reversible aggregates consisting of self-assembled monomeric monoclonal antibody molecules, particularly at high antibody concentrations. While size exclusion chromatography is generally unable to detect these reversible aggregates, dynamic light scattering may be used to determine their presence. Flux decline during virus filtration may be minimized by pretreating the feed using a membrane adsorber in order to disrupt the reversible aggregates that are present. The formation of reversible aggregates is highly dependent on the monoclonal antibody and the feed conditions. For the pH values investigated here, pretreatment of the feed using a hydrophobic interaction membrane adsorber was the most effective in minimizing flux decline during virus filtration. Ion exchange membranes may also be effective if the monoclonal antibody and membrane are oppositely charged. The effectiveness of ion exchange membrane adsorbers is dependent on solution pH . Size based prefiltration was found to be ineffective at disrupting these reversible aggregates. These results can help guide the development of more effective virus filtration processes for monoclonal antibody production.

Insights into the In-Situ Growth of ZIF-L Metal-Organic Framework on Ultrafiltration Membrane: MOF Growth Kinetics Effects on Water Permeability and Dye Separation Performance

Himangshu Mondal (The University of Alabama)*; Milad Esfahani (The University of Alabama)

Metal-Organic Framework (MOF) functionalized membranes have garnered significant attention for precise gas and liquid separation applications. There is an increased attention to utilizing the MOF as the main selective layer among the functionalized membranes. The in-situ growth method is considered reliable as it eliminates MOF aggregation and uneven distribution issues that lead to defective membranes. In this study, a Zeolitic Imidazolate Framework (ZIF-L) MOF was in-situ synthesized using a solution-based room temperature (SBRT) method on a Polyethersulfone (PES) UF membrane by leveraging the -OH and -NH2 functionalities of the PDA. The MOF growth kinetics (from 30 min to 48 h) revealed that 48 h was the optimal time to achieve a 20 μm thick and compact ZIF-L selective layer. The 24 h and 48 h ZIF-L functionalized membranes exhibited water permeabilities of 24 LMH/Bar and 12 LMH/Bar, respectively. The water permeability measured at different time scales provided valuable insights into the ZIF-L growth characteristics on the PDA-coated PES membrane. The confined growth of ZIF-L within the PDA layer required at least 10 h to increase the water permeability compared to PDA coated membrane. The permeability reached its maximum value at 24 h synthesis time, which was equivalent to the pristine PES UF membrane. After 24 hours, the water permeability of the ZIF-L functionalized membranes decreased as the defects and compactness of the ZIF-L layer improved. The ZIF-L functionalized membranes exhibited no MgSO4 salt selectivity. The Zeta Potential showed both 24 h and 48h ZIF-L functionalized membranes were positively charged till pH 8. When challenged with negatively charged Methyl Orange (MO) and Reactive Orange 16 (RO-16) dyes, the rejection rates for the 48 h functionalized membrane improved to approximately 41% and 44%, respectively compared to 10% rejection in Blank and PDA coated membrane.

Ion Transport in Microporous MFI Zeolite Membranes

Hao Yu (Arizona State University)*; Lance Lai (Arizona State University); Jerry Lin (Arizona State University)

Ion transport in microporous materials, such as zeolite membranes, is vital for energy storage and separation technologies, including lithium extraction from seawater or brine. The fully hydrated Li^+ ion (0.76 nm in water) exceeds the pore size of many zeolites, such as MFI (0.56 nm), posing transport challenges. Partial desolvation is needed for Li^+ diffusion in these pathways.

In bulk liquids, ion transport occurs via vehicular and structural diffusion. Vehicular diffusion involves ions moving with their solvation shell intact, while structural diffusion involves ion migration through solvent exchange. How these mechanisms apply to microporous systems remains unclear. This study examines how ion size, charge density, and hydration energy affect transport in confined pathways.

Defect-free polycrystalline MFI zeolite membranes were synthesized. MFI seeds were coated onto an alumina substrate and subjected to secondary growth. Membranes were tested electrochemically with current-voltage (I-V) curves measured for five 0.1 M electrolyte solutions (LiCl, NaCl, KCl, $MgCl_2$, and $CaCl_2$) at room temperature.

A resistance-in-series model corrected for solution, electrode, and membrane resistance, with adjustments for membrane porosity and tortuosity. Ionic conductivities were 1.94×10^{-5} S/cm for LiCl, 1.73×10^{-5} S/cm for NaCl, 1.45×10^{-5} S/cm for KCl, 2.21×10^{-5} S/cm for $MgCl_2$, and 1.16×10^{-5} S/cm for $CaCl_2$. The trend $Mg^{2+} > Li^+ > Na^+ > K^+ > Ca^{2+}$ aligns with bare ion size, indicating transport efficiency is not solely determined by hydration. Selectivity ratios for Li^+ over other ions were 1.49 for Li^+/K^+, 1.34 for Li^+/Na^+, 1.65 for Li^+/Mg^{2+}, and 3.57 for Li^+/Ca^{2+}. Lower Li^+/Mg^{2+} selectivity in the MFI layer suggests intrinsic factors like ion charge density influence transport. Ions with higher charge densities, such as Mg^{2+} and Li^+, exhibit enhanced transport through confined pathways due to partial dehydration and strong interactions with zeolite pore walls.

New High Permeance, Roll to Roll Gutter Layer for Thin Film Composite Gas Separation Membranes

Michael McGrath (Permeance Limited and Bardo Climate Inc)*; Sam Nesbitt (Permeance Limited); Drew Letiu (Permeance Limited); Matthew Cowan (Permeance Limited); Greg Dwulet (Permeance Limited); Lingxiang Zhu (National Energy Technology Laboratory); David Hopkinson (National Energy Technology Laboratory)

The authors chose not to publish the abstract.

Solvent transport through swollen polymers: where do we go now?

Guy Ramon (Technion - Israel Institute of Technology)*

Understanding how various species transport across polymer films is at the core of designing and fabricating membranes with desirable separation properties. However, there are still fundamental uncertainties - a particular example of which are polyamide films used for desalination where, despite their wide-spread commercial deployment, there is still some debate as to the exact transport mechanisms and, consequently, the ability to correctly model separation performance. This is particularly important as future materials – both the polymers as well as the mixtures to be separated – may possess features that impact the way in which transport occurs.

The aim of the present talk is to first discuss the potential pitfalls emanating from misinterpretation of models and the inherent assumptions built into them, as well as the difficulty in translating molecular-scale information into continuum models. A particular point is the physical nature of transport within a swollen polymer network, and the impact of confinement and molecular interactions on the mode of transport.

Use of various modeling approaches, ranging from molecular to process scales, is essential for completing the picture, while maintaining accessibility - the importance of engineers' ability to utilize 'good-enough' models, implemented into design software, cannot be overstated and should be part of the ultimate goal for modeling membrane transport.

Membrane testing at Air Liquide's Innovation Campus Delaware

David Hasse (Air Liquide)*; Marcos Da Conceicao (Air Liquide); Sandeep Karode (Air Liquide); Ramy Swaiden (Air Liquide); Jonathan Leister (Air Liquide); Jake Stephan (Air Liquide); Greg Myrick (Air Liquide)

Air Liquide has developed 4 pilot skids for membrane bundle testing at the Innovation Campus Delaware. Three can test commercial membrane cartridges, with a demonstration scale skid for light gases. The capabilities and some of the results will be presented.

In 2013, Air Liquide added its first commercial scale bundle testing platform to support work with the cold membrane project. In 2021, Air Liquide launched its Outdoor Membrane Test Platform (OMTP) in Delaware, with operations commencing in 2023. This platform houses three test skids complementing the existing cold membrane test skid located indoors. The OMTP offers a range of testing capabilities, including:

Cold membrane testing with live flue gas
Natural gas testing including water, H2S and hydrocarbon addition
Demonstration scale testing of hydrogen and helium separations

In this talk, we will discuss the testing capabilities at our site for cold membrane, biogas, natural gas and helium/hydrogen recovery. In particular, we will discuss the capabilities of onsite testing with live flue gas and the new natural gas testing platform as it applies to our natural gas upgrading commercial offerings. We will also discuss the technical challenges associated with membrane pilots start-up, open loop vs closed loop systems, key learnings and present topical results.

Ultraselective ammonium recovery from wastewater: membrane processes drive materials design

Oded Nir (Department of Desalination and Water Treatment, Zuckerberg Institute for Water Research, The Jacob Blaustein Institutes for Desert Research, Ben-Gurion University of the Negev, Beersheba 84990, Israel)*

File attached

Self-assembling, charge-containing polymers for highly selective and fouling-resistant membranes

Ayse Asatekin (Tufts)*

Most commercial membranes today are composed of a handful of polymers, developed decades ago. These formulations do not take advantage of the wide parameter space available to polymer scientists and engineers, including the self-organization capabilities of polymeric materials. Our group aims to rationally design polymeric materials that self-assemble into functional, biomimetic nanostructures to achieve enhanced membrane performance, with a particular focus on selectivity and fouling resistance. For example, zwitterionic amphiphilic copolymers (ZACs), comprising hydrophobic and zwitterionic monomers, form self-assembled zwitterionic domains that act as a network of nanochannels for water permeation. The effective pore size of ZAC-based thin film composite membranes can be tuned between 0.9-1.5 nm by controlling polymer chemistry and cross-linking. ZAC-based membranes with smallest pore sizes (~0.9 nm) can exhibit unprecedented selectivity between equally charged ions (e.g. Cl-/F-), arising from zwitterion-ion interactions amplified by nanoconfinement. Amphiphilic polyampholytes (APAs), where hydrophobic, anionic, and cationic monomers form a random/statistical terpolymer that is insoluble in water, can also create similar pore structure with access to a broader array of functional groups lining the effective nanopores of these membranes. We have also shown that Amphiphilic PolyElectrolyte Complexes (APECs), formed from blends or bilayers of oppositely-charged but water-insoluble polyelectrolytes, can also create extremely small pores coupled with controlled surface charge. These polymer structures offer many degrees of freedom for the synthesis of polymeric membranes for challenging separations, from ion separations for water treatment and resource recovery to bioseparations and chemical purification.

Effects of humidity on gas transport phenomena in PEG-based CO2 separation membranes

Ryan Johnson (University of Florida)*; Joshua Moon (University of Florida)

Efficiently separating CO_2 from N_2 in flue gas remains a present-day separation challenge in light of global decarbonization efforts. Poly(ethylene glycol) diacrylate (PEGDA) based membranes are promising for this separation due to their high sorption affinity for CO_2 arising from a high density of ether oxygens. However, flue gas streams are generally saturated with water vapor. This can cause issues for many membranes as water vapor is a plasticizing penetrant which typically decreases diffusion selectivity of membranes by increasing effective fractional free volume (FFV). Despite the importance of understanding the role of humidity on the CO_2 separation performance of rubbery polymer networks such as PEGDA, the molecular origins of mixed gas/vapor separation properties of PEGDA copolymer membranes have not yet been studied in detail.

This study aims to quantify the changes to PEGDA copolymer membrane structure under controlled humidity conditions and the resulting effects on macroscopic gas transport properties. Our current measurements indicate that water vapor acts as an anti-plasticizing agent, decreasing the permeability of N_2 at relative humidities of 25–50%, and as a plasticizer at relative humidities above 75%, increasing N_2 permeability compared to dry conditions, independent of the crosslinking density of these membranes. CO_2 permeabilities in membranes with 20-80 wt% crosslinker are constant at relative humidities up to 75% where the CO_2 permeability decreases by up to 25%. This presentation will further discuss how gas permeabilities are affected by humidity-induced changes in membrane FFV. Effective FFV values in water-plasticized membranes can be evaluated through combining isotherms from water vapor sorption and water vapor dilatometry measurements. The fundamental insights from this study will ultimately be useful for elucidating the effects of complex gas/vapor feed mixtures on structure-property transport relationships for rubbery membranes.

Patterned anion exchange membranes for abiotic, medically implanted glucose consuming fuel cells

Thomas DosSantos DiSorbo (University of Colorado Denver | Anschutz)*; John Pellegrino (University of Colorado Boulder); Richard Weir (University of Colorado Denver | Anschutz)

Many implantable medical device technologies require a safe and efficacious electrical power source to function over a prolonged period of use. To increase device longevity and reduce the complications of battery replacement, energy harvesting technologies like abiotic glucose fuel cells (AGFCs)—which generate electrical power by catalyzing the oxidation of the blood's glucose to create an electrical current—are under investigation as viable biocompatible energy sources. Our current focus is studying patterned, commercial anion exchange membranes (AEM) for AGFCs. Under physiological pH, glucose oxidation kinetics are slow and we are seeking any design advantage to increase the power density. AEMs are an important component of AGFCs, facilitating selective ionic transfer between the cell cathode and anode, and patterning may provide higher ion mass transfer.

We report benchtop testing of several membrane electrode assemblies (MEAs) to study the advantages of using a patterned AEM in a Pt-catalyzed AGFC. Patterned AEMs have a larger surface area and may facilitate ionic transport, as well as improve catalyst contact. The superior conductivity of patterned membranes have been observed in proton exchange membranes (PEMs) for hydrogen fuel cells, but the technology is novel to AEMs.

We evaluated AEMs imprinted with both micron and sub micron sized patterns across varied, non physiological, fuel mixtures of glucose and KOH in order to obtain higher power outputs for these initial studies. The micron AEM patterning yielded a statistically significant improvement to power output in the operational range for most fuel compositions when compared to an unpatterned control. AEM orientation with respect to facing anode versus cathode was investigated to provide potential mechanistic insights. In general, the micron patterned AEM enhanced performance in both orientations.

Carbon Molecular Sieves Prepared by Moderate-Temperature Carbonization of Phosphoric Acid-Doped Polyimide for O2/N2 Separation

Yang Jiao (University at Buffalo)*; Haiqing Lin (University at Buffalo)

Polyimide-derived carbon molecular sieve (CMS) membranes have been extensively explored to obtain hierarchical free volumes including ultramicropores (< 1 nm) with strong size sieving ability and micropores (1 - 2 nm) with high gas diffusivity for O_2/N_2 separation. However, the carbonization temperature needs to be very high (800 ℃ or above) to achieve the desired ultramicropores (< 3.64 Å) and sufficient O_2/N_2 selectivity. Herein we demonstrate that a highly permeable polyimide (6FDA-DAM) can be doped with phosphoric acid (PA) before carbonization at moderate temperatures to achieve desired ultramicropores. PA forms hydrogen bonds with the 6FDA-DAM and can be degraded at much lower temperatures to manipulate the free volumes. The effect of PA doping level and carbonization temperature on physical and O_2/N_2 separation properties is thoroughly investigated. Increasing the carbonization temperature decreases the d-spacing of the PA-doped 6FDA-DAM. For instance, the carbonization 600 oC decreases the d-spacing from 5.9 to 3.9 Å, and an increase to 900 oC further decreases to 3.5 Å. The CMS derived from the PA doping level of 0.25 (defined as the molar ratio of PA to the polymer repeating units) exhibits the best improvement in O_2/N_2 separation properties. For example, when carbonized at 700 oC, the CMS of 6FDA-DAM0.25 (6FDA-DAM0.25@700) shows O2 permeability of 300 Barrer and O2/N2 selectivity of 11, surpassing Robeson's upper bound. By contrast, 6FDA-DAM0.25 and 6FDA-DAM@700 show O2/N2 selectivity of only 3.7 and 9.3, respectively, and O2 permeability of 120 and 330, respectively. The doping level and pyrolyze temperature are optimized to obtain the best O2/N2 separation properties. Our approach provides a simple and efficient method for tuning the pore size of CMS materials while retaining their flexibility and processability, thereby enhancing their size-sieving capability for precise gas separation applications.

Extracellular polymeric substances attributes to biofouling of anaerobic membrane bioreactor: Adhesion and viscoelastic properties

Kbrom mearg Haile (Benguroin university of the negev ,Israel)*

Membrane fouling limits the robust and continuous operation of anaerobic membrane bioreactor (AnMBR), caused by the mixed liquor suspended solids (MLSS) characteristics formed via aggregation of flocs and microbial scaffold. self-produced extracellular polymeric substances (EPS), which dictates the floc's integrity. The adhesion of EPS to the membrane versus their role in forming firm, elastic, and mechanically stable flocs under hydraulic shear is critical for minimizing interactions between EPS and colloids, originating from the MLSS flocs, with the membrane. This study explores the effect of coagulant ($FeCl3$) on MLSS flocs properties, EPS adhesion and viscoelasticity, and membrane fouling propensity.

A flat-sheet 200 kDa PES membranes ($352cm^2$) were immersed in an AnMBR treating municipal wastewater. System conditions (temperature, pH, OLR, biogas, and HRT) were controlled. TMP changes were monitored under two operating conditions: baseline (no FeCl3) and with FeCl3 addition. At the end, EPS from MLSS and fouled membranes were analyzed for protein, polysaccharides, and DOC, and linked to membrane fouling. EPS adhesion and viscoelasticity were studied using QCM-D with a PES-coated gold sensor. Frequency/dissipation shifts were modeled with the Voigt-based model (Dfind software) to calculate EPS layer thickness, shear modulus, and viscosity.

The QCM-D analysis indicates a greater decrease in frequency shift in the elevated fouling scenarios, likely due to a lower shear viscosity and modulus of the EPS layer, with an increase in its hydrated thickness and fluidity ($\Delta D/\Delta f$). Further sludge rheology and dewaterability analysis showed improved mechanical stability and solids dewaterability.The biofouling layer is further characterized using a confocal laser scanning microscope (CLSM) and scanning electron microscope (SEM) and thinner biofouling layer was observed when operated with coagulant addition compared to the baseline experiment, due to enhanced flocs integrity.

Active Fouling Removal from Membranes with Interdigitated Electrodes

Bezawit Getachew (Rice University)*

Electrically conductive membranes offer a promising alternative to traditional cleaning methods by reducing or eliminating the need for additional chemicals and lowering the frequency of taking membrane modules offline for cleaning. Two outstanding challenges in active fouling removal with electrically conductive membranes are simple ways to incorporate a counter electrode in membrane modules and the impact of the conductive material on membrane transport properties. This study employed patterned electrically conductive membranes with interdigitated working and counter electrodes on the membrane surface, as a way to limit negative impacts on flux and easily incorporate two electrodes on a membrane. A systematic investigation of design parameters, including electrode widths (1–2 mm) and spacings (1–3 mm), showed self-cleaning efficiencies of sodium alginate (SA)-fouled membranes ranging from 20–60%, with improved performance achieved through thicker, narrower-spaced conductive lines.

The relationship between membrane fouling index (MFI) and electrochemical self-cleaning was also explored. Variations in foulant concentration, represented by MFI values for SA concentrations of 1, 3, and 5 mg·L-1, revealed that lower fouling rates (lower MFI values) corresponded to reduced self-cleaning efficiency. ToF-SIMS analysis and depth profiling indicated distinct fouling distributions: membranes fouled with 5 mg·L-1 SA had fouling concentrated on the surface, while membranes fouled with 1 mg·L-1 SA exhibited fouling within the pores. These differences in fouling distribution likely contribute to the observed variations in self-cleaning efficiency, providing valuable insights for the optimization of interdigitated-patterned electrically conductive membrane designs.

Nanofiltration using Polysulfone Membranes

Jian Tan (Pall Corporation)*; Varun Vakharia (Pall Corporation)

Defect control is essential to leading-edge semiconductor industry. Nano filtration has been effective at reducing defects at low cost. In this study, we focus on filters that consist of polysulfone (PSF) membrane and ion exchange membrane (IEM) for effective removal of both particle and metal ion contaminations. In the real application case, either one filter containing dual layer membranes (PSF + IEM) or two separate filters (PSF filter + IEM filter) can be installed to remove particles and metal ions from chemicals or waters used in semiconductor manufacturing, depending on the conditions and requirements. Polysulfone polymer membrane is a highly asymmetric structure that allows superior flow rate and fine particle retention. It is pleated and packed to form filters for maximum use of its area. The pleats typically include membrane and support/drainage materials, and are enveloped by endcaps, outer cage and internal support core where needed. With optimal filter design, the best filter has a rating of 2 nm while holding a low pressure drop. IEM is typically modified with negatively charged functional groups to remove positively charged metal ions. The leading IEM product can effectively remove 99% trace metal ions from solutions. This study will give an overview of the issues at end users and discuss how advanced filtration technology help solve the issues and enable semiconductor manufacturing.

Looking beyond size and charge: A roadmap to ion separations for the global energy transition

Ryan Kingsbury (Princeton University)*

Ion separations (and separation materials such as membranes) are a crucial enabling technology for the energy transition. First, ion conducting membranes are ubiquitous components of electrochemical devices ranging from redox flow batteries and fuel cells to processes for decarbonized chemical manufacturing. Second, many of the minerals deemed “critical” by the Dept. of Energy and essential to renewable energy technologies, including Li, Ni, Mn, and Co, occur as dissolved ions that can potentially be recovered from natural and man-made brines.

The multitude of emerging applications in these two areas has led to an explosion in the number of specific separation challenges (e.g., combination of target ion, source water, competing background ions, desired selectivity, etc.) that are potentially of interest. Notably, many emerging separations require a degree of “specific ion selectivity”—the ability to differentiate ions with the same charge such as Li+ and Na+—that far exceeds the capabilities of state-of-the-art materials such as reverse osmosis and ion exchange membranes.

In this presentation, I will argue that developing selective membranes to address this huge variety of emerging ion separation problems requires a roadmap—analogous to the periodic table—that facilitates a more systematic understanding of separation mechanisms. I will present a comprehensive analysis of the properties of ion species relevant to emerging applications in electrochemistry, critical mineral and resource recovery, revealing fundamental asymmetries between cations and anions, and discuss their implications for separation design. I will highlight the importance of understanding the electronic structure (i.e., the shape and character of electron clouds) of ionic species, and close by presenting a "periodic table" that rationalizes ion selectivity trends reported in a variety of membrane and sorption processes.

Development of highly hydrophobic self-heating Fe3O4-PVDF membranes

Amanda Loreti Hupsel (COPPE/PEQ/UFRJ)*; Cristiano Piacsek Borges (COPPE/PEQ/UFRJ); Fabiana Valéria da Fonseca (EQ/UFRJ)

Magnetite nanoparticles (Fe3O4-NPS) are promising materials that exhibit properties such as superparamagnetic comportment and high coercivity. Membranes containing Fe3O4-NPS may have self-heating properties, which is very interesting for use in membrane distillation (MD), aiding in reducing the phenomenon of temperature polarization. However, the hydrophilic nature of these nanoparticles may reduce the contact angle of the MD membranes, which is undesirable. The objective of this study was to prepare flat-sheet PVDF membranes containing Fe3O4-NPS with a highly hydrophobic surface and self-heating properties, aiming for use in MD. PVDF membranes were prepared by non-solvent-induced phase separation (NIPS) with 20% wt. of polymer and 5% wt. of oleic acid (OA) as an additive, and water or ethanol as coagulation baths. An 1% wt. load of pristine or oleic-acid coated Fe3O4-NPS was added to the dope solutions. Pristine PVDF membranes and membranes with only OA or Fe3O4-NPS as additives were also prepared to better understand the individual effects of each additive. The different NIPS conditions resulted in membranes with diverse morphologies. The use of OA and ethanol coagulation bath contributed to the synthesis of porous membranes with hydrophobic surfaces. The P-20-OA-NOA-E membrane prepared in ethanol coagulation bath with oleic-acid-modified Fe3O4-NPS exhibited a bottom-layer water contact angle (WCA) of 103° and a superhydrophobic top-layer WCA of 150°C, with better dispersion of the Fe3O4-NPS on the membrane matrix compared to the membrane with unmodified NPS. The membrane exhibited a unique Janus-like morphology, with a rougher top surface, which contributed to its higher hydrophobicity. The membranes were able to generate heat under a low-power 12V induction heating system, which shows that they have self-heating properties, demonstrating their potential for use in MD.

Innovative zeolite membrane reactors for Dimethyl ether production by MeOH dehydration

Giuseppe Barbieri (Consiglio Nazionale delle Ricerche - Istituto per la Tecnologia delle Membrane)*; Elisa Avruscio (Consiglio Nazionale delle Ricerche - Istituto per la Tecnologia delle Membrane); Abdu W. Sabir (Chung-ang University); Pyung Soo Lee (Chung-ang University); Adele Brunetti (Consiglio Nazionale delle Ricerche - Istituto per la Tecnologia delle Membrane)

Dimethyl ether (DME) is a promising e-fuel due to its clean combustion and reduced emissions. It is produced via methanol dehydration using solid acid catalysts like zeolites. Zeolite catalysts, particularly ZSM-5 and BEA, are preferred over Al2O3 due to their better water stability, although they can promote unwanted reactions above 250°C, including coke formation.

In this study, a zeolite membrane reactor (MR) was used as a contactor for DME production by methanol dehydration, employing innovative home-made BEA and ZSM-5 membranes [1]. The MR operated in a steady-state continuous "through flow" configuration, allowing the entire feed to pass through the membrane. This approach aimed to promote the removal of water from catalytic sites, reducing catalyst deactivation typically observed in traditional catalytic beds. The results were evaluated in terms of methanol conversion, DME yield, selectivity, and stability over time. The performance of each MR was analyzed as a function of temperature (200-300°C), WHSV (2.6-21.1 gMeOH gCatalyst-1 h-1), and feed composition (80-100% mol MeOH). Both BEA and ZSM-5 membrane reactors exhibited full DME selectivity under all investigated conditions. BEA membranes showed better performance, achieving 88% conversion at 200-220°C and 3.5 h-1, close to the equilibrium value. A slight drop in conversion to 80% was observed at 260°C, likely due to thermodynamic effects. ZSM-5 membrane reactors consistently showed lower performance, attributed to their smaller pores and fewer acid sites. BEA membranes, with larger pores and more acid sites, exhibited better performance but also more non-zeolitic defects due to the low Si/Al ratio. Stability tests on BEA membranes (150 h at 3.5 h-1, 260°C) showed a decrease in methanol conversion, which was fully restored after thermal treatment at 550°C for 8 hours.

Acknowledgements - HydroMine project co-funded by the European Union's Horizon RFCS-2022 - GA 101112629 is gratefully acknowledged.

Characterization, Performance Evaluation, and Toxicological assessment of Polysulfone-Sulfonated Polyether Ether Ketone Membranes for Water separation Applications

Muhammad Usman Yousaf (University of Kentucky)*; Lucca Cortarelli (Department of Plant and Soil Sciences, Martin-Gatton College of Agriculture, Food and Environment, University of Kentucky, Lexington, KY, USA); Nerissa Kiprotich (Department of Chemical and Materials Engineering, Stanley and Karen Pigman College of Engineering, University of Kentucky, Lexington, KY, USA); Nirupam Aich (Department of Civil and Environmental Engineering, University of Nebraska – Lincoln, Lincoln, NE, USA); Olga V. Tsyusko (Department of Plant and Soil Sciences, Martin-Gatton College of Agriculture, Food and Environment, University of Kentucky, Lexington, KY, USA); Isabel Escobar (Department of Chemical and Materials Engineering, Stanley and Karen Pigman College of Engineering, University of Kentucky, Lexington, KY, USA)

Removal of charged organic dyes from aqueous solutions using polymeric membrane remains challenging due to their small molecular size and high solubility. In this study, polysulfone (PSf) and sulfonated polyether ether ketone (SPEEK)-based membranes were fabricated via the non-solvent induced phase separation process (NIPS) using N-Methyl-2-Pyrrolidone (NMP) as solvent and water as non-solvent. Variations in polymer concentration were investigated to identify impacts on membrane structure and performance. Additionally, annealing was investigated to assess if heat treatment of the membrane prior to operation could reduce toxic solvent leaching since a key focus of this work was membrane fabrication under a Safe(r)-by-Design framework targeting functional performance and reduced environmental toxicity.

Membranes were characterized structurally and morphologically followed by toxicity assessment before and after filtration with and without annealing at various pH to test for potential leaching of any trapped solvent from pores. Additionally, membranes were also characterized with respect to operational performance using binary mixtures of cationic and anionic dyes. Membranes demonstrated selective filtration behavior in which cationic dyes were preferentially rejected due to size exclusion and electrostatic interaction. In a binary dye filtration experiment, Methylene Blue, a cationic dye, exhibited above 98% rejection, while, Acid Orange 2, an anionic dye, exhibited gradually increasing rejection due to adsorption to the membrane and charge modification effects. Permeability decreased over time, likely due to surface charge changes and the formation of aggregates caused by interactions between oppositely charged dyes. Toxicity assessments using a model organism, a nematode C. elegans, showed that annealing reduced solvent leaching and thus permeate toxicity, particularly at neutral pH, by facilitating trapped solvent release prior to membrane use.

Evaluation of hollow fiber membrane contactor reactor in reactive extraction for biodiesel production

Frederico Kronemberger (PEQ/COPPE/UFRJ)*; Nevardo Yaya (PEQ/COPPE/UFRJ); Claudio Habert (PEQ/COPPE/UFRJ)

Membrane contactors enable non-dispersive contact between fluid phases using a membrane as an interphase barrier. This technology offers advantages over conventional dispersed-phase contactors, including high interfacial area, low energy consumption, and ease of scale-up. Biodiesel is industrially produced via methanolysis, which involves transesterification of oils with methanol using a basic catalyst. Conventional processes rely on stirred reactors to overcome phase immiscibility, but face limitations due to equilibrium constraints, high raw material costs, and complex purification steps. Process intensification strategies, such as membrane reactors, offer a means to improve the efficiency by integrating the reaction and separation in a single unit. This study explores HFMC technology for biodiesel production. Custom polypropylene HFMCs were developed for the methanolysis of soybean oil and used as hollow fiber membrane contactor reactors (HFMCRs) in a reactive extraction process. A mathematical model was created to predict reactor performance, incorporating mass transfer limitations and flux through the membrane. Experimental validation assessed the effects of oil flow rate (OFR), methanol-to-oil molar ratio (MR), and packing density (PD) on triglyceride conversion and biodiesel (FAME) yield.

The results showed that a countercurrent HFMCR configuration enables simultaneous reaction and separation, where methanol extracts glycerin from the biodiesel-rich phase, overcoming equilibrium constraints. This reduces the need for separate glycerol removal, improving productivity, and simplifying downstream purification. The developed model accurately described experimental data, estimating the key parameters for the shell-side mass transfer. A conversion of 35% and a yield of 20% were achieved with PD = 30%, MR = 4:1, and OFR = 0.4 L/h. This study demonstrates the potential of HFMCs for the biodiesel production, reducing processing time and downstream separation burdens.

Scalable Synthesis of Electrically Conductive Polyethersulfone (PES) Membranes

Abdelrahman Awad (McMaster University)*; Nathan Mullins (McMaster University); Charles De lannoy (McMaster University)

Several challenges are inherent in the synthesis and scalability of electrically conductive membranes (ECMs). ECMs are prepared by pressure coating of polymer membranes with conductive nanomaterials, but this coating synthesis greatly reduces the membrane permeability as conductive materials can block the membrane pores. Another limitation is the manufacturing scalability of ECMs, as vacuum deposition is not a scalable roll-to-roll method. In this research, highly conductive membranes were fabricated by a facile sequential casting process that is both scalable and led to a controllable water permeability. A polymer dope of polyethersulfone (PES 15 wt%) and polyvinylpyrrolidone (5wt%) in N-Methyl-2-pyrrolidone (NMP) was cast on a glass plate to form a support PES layer. Subsequently, carbon nanotubes (CNTs) dissolved in NMP and propylene glycol were cast atop a wet polymer dope to form a thin conducive skin layer. The two-layer cast solution was immersed in a coagulation bath to make the ECMs by the nonsolvent-induced phase separation. Fabricated membranes were thoroughly characterized for their conventional properties (e.g. permeability, rejection) and their electrochemical properties. Control membranes were prepared by vacuum deposition of CNTs on pre-made PES substrates and compared to the sequential casting-based ECMs. ECMs prepared by sequential casting could be made with high electrical conductivities (80-120 Ω) and achieved 98% electrochemical removal of methyl orange at -3V. Defect-free conductive PES membranes can be formed by controlling the solvent-nonsolvent diffusion rates, which was achieved by manipulating the viscosities of CNT solvent and the coagulation bath. Fabricated ECMs had smaller pore sizes, which led to a reduction in water permeability as compared to PES control membranes (~ 266 L. m-2. h-1.bar-1). However, ECMs synthesized by sequential casting exhibited a 40% higher water permeability than PES/CNTs membranes prepared by vacuum deposition.

Engineering ion transport in polymer membranes for water purification and energy applications

Geoffrey Geise (University of Virginia)*

Providing sustainable supplies of purified water and energy is a critical global challenge for the future, and polymer membranes will play a key role in addressing these clear and pressing global needs for water and energy. Polymer membrane-based processes dominate the desalination market because they are more energy efficient than thermal desalination processes, and polymer membranes are crucial components in several rapidly developing power generation, energy storage, and specific ion capture/recovery applications that rely on membranes to control, selectively, rates ion transport. Much remains unknown about the influence of polymer structure and polymer processing/manufacturing on intrinsic transport properties, and these relationships must be developed to design next generation polymer membrane materials. This presentation will provide an overview of research focused on characterization of small molecule transport (e.g., water, ions, small molecule organics) in hydrated or organic solvent swollen polymer membranes that are of interest for water purification and/or clean energy applications. Additionally, the presentation will discuss the use of ion transport characterization and microwave dielectric relaxation spectroscopy capabilities to uncover structure-property relationships that lead to materials engineering principles and guidelines for preparing next generation membranes for separations.

Overcoming the Conductance versus Crossover Trade-off in State-of-the-Art Proton Exchange Fuel-Cell Membranes by Incorporating Atomically Thin Chemical Vapor Deposition Graphene

Nicole Moehring (Bosch USA); Abdul Mansoor Basha (University of Calgary); Pavan Chaturvedi (Vanderbilt Univeristy)*; Thomas Knight (UT Knoxville); Xiaozong Fan (Vanderbilt University); Peter Pintauro (Vanderbilt University); Michael Boutilier (University of Western Ontario); Kunal Karan (University of Calgary); Piran Kidambi (Vanderbilt University)

Permeance–selectivity trade-offs are inherent to polymeric membranes. In fuel cells, thinner proton exchange membranes (PEMs) could enable higher proton conductance and increased power density with lower area-specific resistance (ASR), smaller ohmic losses, and lower ionomer cost. However, reducing thickness is accompanied by an increase in undesired species crossover harming performance and long-term efficiency. Here, we show that incorporating atomically thin monolayer graphene synthesized via scalable chemical vapor deposition (CVD) and tunable defect density into PEMs (Nafion, ~5–25 μm thick) can allow for reduced H2 crossover (~34–78% of Nafion of a similar thickness) while maintaining adequate areal proton conductance for applications (>4 S cm–2). In contrast to most prior work using >50 μm symmetric Nafion sandwich structures, we elucidate the interplay of graphene defect density and Nafion proton transport resistance on the performance of Nafion|graphene composite membranes and find high-quality low-defect density CVD graphene (G) supported on Nafion 211 (~25 μm); i.e., N211|G has a high areal proton conductance (~6.1 S cm–2) and the lowest H2 crossover (~0.7 mA cm–2). Fully functional centimeter-scale N211|G fuel-cell membranes demonstrate performance comparable to that of state-of-the-art Nafion N211 at room temperature as well as standard operating conditions (~80 °C, ~150–250 kPa-abs) with H2/air (power density ~0.57–0.63 W cm–2) and H2/O2 feed (power density ~1.4–1.62 W cm–2) and markedly reduced H2 crossover (~53–57%).

Reference: Moehring et al., Nano Lett. 2025, 25, 3, 1165–1176

Beyond Power: CO2 Capture from Industrial Sources with Polaris™ Membranes

Richard Baker (Membrane Technology & Research Inc.); Brice Freeman (Membrane Technology & Research Inc.); Pingjiao Hao (Membrane Technology & Research Inc.); Jay Kniep (Membrane Technology & Research Inc.)*; Tim Merkel (Membrane Technology & Research Inc.); Robert West (Membrane Technology & Research Inc.); Kevin Lauzze (Sargent & Lundy); Dana Pierik (Sargent & Lundy)

Carbon capture from industrial point sources is an important part of any carbon management policy that has a target of reaching net zero emissions. Many persistent industrial sources, such as cement and steel production, generate CO2 emissions even in the absence of fossil fuel use. However, today's capture options tend to be complicated and expensive. For these reasons, DOE has funded development of a suite of capture technologies based on advanced solvents, sorbents or membranes.

Membrane processes offer significant advantages for CO2 capture compared to alternatives. Working with DOE, MTR has developed the industry-leading Polaris™ membrane, innovative process designs, and low pressure drop membrane modules for carbon capture applications. This work has advanced membrane carbon capture innovations from concept through large pilot demonstration and FEED studies.

To date, most R&D has centered on capturing CO2 emitted from power plants, as they are the largest point source emitters globally. However, industrial plants also emit large amounts of CO2 and are potentially easier targets for a CO2 capture process because the CO2 in their flue gas tends to be more concentrated. A study by SINTEF examined a capture process modeled on the MTR design and found a capture cost for cement plant flue gas that was 9% lower than capture using the reference monoethanolamine (MEA) process. More recently, DOE completed a study of capture costs from a cement plant using CANSOLV® amine. The cost of capture for the MTR process was found to be 10% – 30% lower than CANSOLV with larger savings at higher feed CO2 contents (range studied was 14.5 to 31 mol% CO2).

In this talk, a detailed analysis of membrane-based systems for the capture of the CO2 emitted from the cement production process will be presented. Process designs, capture costs and economics, and an update on a current small pilot field test project at a cement plant in Michigan will be discussed.

Comparison of classical hydrodynamic models of transport through porous membranes

Minhao Xiao (UCLA)*

Herein, we compare three hydrodynamic models for the water permeance (Hagen-Poiseuille, Kozeny-Carman, and Happel cell) and solute rejection (Ferry, Ferry-Renkin, and Zeman-Wales) of porous microfiltration (MF) and ultrafiltration (UF) membranes. We fabricate a series of porous membranes via nonsolvent induced phase separation (NIPS) comprising different concentrations of polysulfone (PSU), polyvinylidene fluoride (PVDF), and polyethersulfone (PES). Analysis of scanning electron microscope (SEM) images enabled quantification of surface porosity, mean pore size, and skin layer thickness of dry membranes. Water uptake and membrane swelling measurements yield the dry-to-wet conversion factor that allows one to reproduce the fully wetted membrane pore morphology from dry membrane SEM images. Dead-end stirred-cell filtration data establishes the observed water permeance and solute rejection by the membranes. After applying the dry-to-wet conversion factors, Ferry's mechanical sieving model predictions fit experimental rejection data almost perfectly. The Kozeny-Carman permeability model agreed best with experimental data in the lower water permeance range (10 to 200 lmh/bar), whereas the Hagen-Poiseuille made better predictions in the higher permeance range (200 to 3,500 lmh/bar). That said, Happel's cell model predictions most closely matched experimental water permeance data over the entire range. A new equation of porous membrane water permeance is proposed based on best-fit results from the membranes tested herein, while the original Ferry mechanical sieving model works nearly perfectly for uncharged solutes using SEM-derived mean pore size. Our results clearly illustrate the role of porous membrane swelling in aqueous filtration applications and must always be considered in membrane characterization studies.

Engineering thin film composite gas separation membranes using crosslinked nanoporous macrocycles

Joshua Moon (University of Florida)*; Woochul Kwak (University of Florida); Madeline Howell (University of Florida); Kaitlin Slicker (University of Florida); Ryan Johnson (University of Florida); Austin Evans (University of Florida)

Challenges with reliably manipulating size-sieving in conventional membranes has driven research toward nanoporous covalent organic and metal-organic frameworks. While COFs and MOFs can yield rigid, size-sieving nanopores using a fixed bond framework, their reversible chemistry often leads to unavoidable defects, grain boundaries, and disordered packing. We are developing an analogous strategy to access precise nanostructures via 2D polymerization and assembly of organic macrocycles, which are small cyclic molecules containing intrinsic nanocavities. Macrocyclic membranes are particularly attractive as gas separation membranes because (1) they are exceptionally thin (sub-100 nm), (2) they are constructed through chemically robust covalent amide bonds, and (3) they are densely crosslinked which potentially reduces their susceptibility to plasticization.

This presentation will discuss our current NSF supported work on fabricating thin film composite (TFC) gas separation membranes using macrocycles as core building blocks. Prior work from the Evans group has demonstrated that calixarene macrocycle membranes exhibit solvent transport rates that are an order of magnitude faster than state-of-the-art commercial membranes. In our present work, polyamide thin films are formed via interfacial polymerization using cyclam (cyclic amine) macrocycles with film thicknesses controllable on the order of hundreds of nanometers. TFC membranes are fabricated using a porous poly(ether sulfone) support and poly(dimethylsiloxane) gutter layer. Separation performance of macrocyclic polyamide membranes is compared with that of conventional aromatic polyamide films.

Pressure-dependent gas permeances (He, N2, CO2) of macrocyclic films determined from a series resistance model provide further insights into the degree to which transport follows a solution-diffusion vs. Knudsen diffusion mechanism.

Ceramic Ion Exchange Membrane Operational Stability in Organic Solvents

Stephanie Candelaria (Membrion)*; Ryan Flores (Membrion); Unnati Rao (Membrion); Emily Rabe (Membrion); Greg Newbloom (Membrion)

Membrion has developed novel nanoporous ceramic ion exchange membranes (IEMs) using our patented sol-gel process, which enables large-scale roll-to-roll production at ambient temperatures. We produce both cation exchange membranes (CEMs) and anion exchange membranes (AEMs) with highly tunable physical properties, including membrane thickness, pore structure, and surface chemistry. Our ceramic IEMs offer several key advantages over traditional polymer IEMs: high performance at low pH, chemical resistance to oxidizers, and a rigid structure that minimizes swelling.

Membrion's ceramic IEMs were originally developed for aqueous-based electrochemical applications, including redox flow batteries and electrodialysis. However, new applications are emerging that require ionic separation in organic solvents. Traditional polymer IEMs can struggle in such conditions, where swelling can limit mechanical stability. In contrast, Membrion's ceramic IEMs have shown remarkable stability in a variety of organic solvents.

Small-angle X-ray scattering (SAXS) studies investigating nanostructure differences in aqueous and non-aqueous environments show that Membrion's ceramic IEM pore size is unchanged, whereas Nafion's pore size increases significantly in the organic solvent dimethylacetamide (DMA). When loaded into an electrochemical cell, Nafion showed swelling and deformation with operation in DMA, whereas the lack of swelling for Membrion's ceramic IEMs suggests long-term durability. Additionally, Membrion's ceramic IEMs demonstrated excellent stability and performance in other applications using organic solvents including non-aqueous redox flow batteries and metal removal using electrodialysis.

Sterically stable nanocrystal cellulose (SNCC) modified microfiltration membranes for the dual-enhanced NOM fouling and biofouling resistance

Ji Qin (University of Minnesota)*; Boya Xiong (University of Minnesota)

Natural organic matter (NOM) fouling and biofilm formation are critical challenges in membrane separation technologies for water treatment, limiting their efficiency and operational lifespan. Nanocrystalline cellulose (NCC) is an emerging, renewable, and low-toxicity nanomaterial that enhances the water treatment membrane antifouling properties due to its superior hydrophilicity but lacks stability, antibacterial properties, and anti-biofouling function. Here, we developed a novel, stable self-assembled sterically stabilized nanocrystalline cellulose (SNCC) layer with superior anti-NOM fouling and antibiofilm, functionalized on commercial microfiltration (MF) membrane. SNCC was derived from partially oxidized dialdehyde cellulose microfibers with a high content of accessible aldehyde groups at amorphous ends, enabling in-situ self-assembly via hemiacetal bonding and forming a stable layer on polydopamine (PDA)-coated MF membranes. Comprehensive surface chemistry and morphology characterization confirmed the formation of a stable, and hydrophilic SNCC self-assembled layer. Filtration tests showed that SNCC-modified membranes improved permeability due to enhanced hydrophilicity. Antibacterial analysis shows that SNCC-modified membranes exhibited significant Escherichia coli and Bacillus subtilis inactivation compared to control. Antifouling tests using NOM solution revealed 7 times less increase in transmembrane pressure (constant flux) for SNCC-modified membranes compared to control. In repeated biofilm fouling and cleaning cycles using crossflow filtration, SNCC-modified membranes achieved significantly higher flux recovery (88-100%) compared to control (33-77%). Confocal laser scanning microscopy further confirmed the effective inhibition of biofilm formation on the SNCC-modified membrane surface. This study provides a novel self-assembled SNCC-modified membrane as effective fouling resistance and biofilm control for sustainable water treatment applications.

Fabrication and Performance of Bipolar Membranes with 3D Junctions

Michael McGrath (Permeance Limited and Bardo Climate Inc)*; Zhezhou Yang (Vanderbilt University); Ryszard Wycisk (Bardo Climate Inc); Peter Pintauro (Bardo Climate Inc)

Bipolar membranes (BPM), which are laminates of anion-exchange and cation-exchange membranes, are used in a variety of industrial, environmental, and climate-related electrochemical applications. Present-day BPMs, with a planar 2D bipolar junction (where the anion-exchange and cation-exchange membranes meet and where water dissociates into H+ and OH- during reverse bias operation) are rated/advertised to operate at current densities <0.1 A/cm2. Preliminary techno-economic modeling suggests that Bardo's next-generation BPMs, operating at 1 A/cm2, can reduce the size of equipment and levelized cost of ED CO2 removal by >50%. Similarly, the ability of Bardo BPMs to operate in water splitting and water generation (forward bias) modes makes them excellent candidates for acid/base redox flow batteries.

In this talk the method of fabricating 3D junction bipolar membranes will be described and the performance of BPMs will be discussed for different anion-exchange polymers (quaternized polyphenylene oxide, AemionTM from Ionomr Innovations, Inc. or PiperION-A TP-85 from Versogen) and cation-exchange polymers (sulfonated poly(ether ether ketone) or Nafion from Chemours). The use of different catalyst particles in the junction layer will also be presented. Current density vs. voltage polarization curves will be shown for BPMs operating in reverse and forward bias modes. The results of extended operation experiments at a constant current density and elevated temperature (50-60°C) will also be presented.

As an example, the long-term stability of a Nafion:TP-85 membrane containing graphene oxide as the catalyst during water splitting at a constant current density of 0.5 A/cm2 and a solution temperature of 45- 50oC is shown in Figure 1. The 3D junction BPM was stable for over 214 hours at 0.5 A/cm2. Water splitting occurred at a low transmembrane voltage drop of only 1.26V.

Acknowledgement - This study was funded by the ARPA-E, under Grant No. DE-AR0001035.

Scalable Manufacture of Loose-Nanofiltration Polymeric Membranes with Green Solvents using Slot Die Coating on a Roll-to-Roll System: Experimental and Computational Insights

Kwangjun Jung (Georgia Institute of Technology)*; David Lu (University of Kentucky); Isabel Escobar (University of Kentucky); Tequila Harris (Georgia Institute of Technology)

Traditional polymeric membrane fabrication relies on petroleum-derived solvents, which poses ecological hazards. In response, this study explored a greener alternative for the filtration of these contaminants by utilizing Polysulfone (Psf) dissolved in co-solvents PolarClean and gamma-Valerolactone (GVL), Psf-PolarClean-GVL, which has shown promising results for removing contaminants, such as polyfluoroalkyl substances (PFAS). The membranes were fabricated using a roll-to-roll (R2R) slot die coating process, followed by phase inversion in a deionized water bath, producing defect-free membranes with thickness ranging from 70 to 200 um. The impact of different support layer configurations on membrane processing and performance was also explored. Filtration and permeability tests using different configurations of fabricated membranes using bovine serum albumin (BSA) as a model solute for ultrafiltration (UF) and nanofiltration (UF) demonstrated a maximum BSA rejection of 99.2% ± 1.31% and mean permeability of 70.5 ± 8.33 LMH/bar.

To enhance scalability and optimize the process parameters, a 2-dimensional multiphase computational fluid dynamics (CFD) model, based on volume of fluids (VOF), was developed using OpenFOAM v2012 to predict the operational limits of processing this dope solution in slot die coating process. Experimentally, periodic voids were observed at solid liquid interface of fabricated membrane through cross-sectional SEM image as shown in Figure 1 (a). The model captured this phenomenon as shown in Figure 1 (c). The model was validated against experimental results showing less than 15 % error. The model was also compared with the previous COMSOL-based model from existing literature, showing less than 10% error, confirming its robustness. These findings provide critical insights into the scalable fabrication of high-performance loose-nanofiltration membranes using green solvents, advancing environmentally friendly water filtration technologies.

Bioinspired Zwitterionic Membrane Functionalized with WS2 nanosheets for Dye Removal through Forward Osmosis Systems

Luis Bermúdez (University of Puerto Rico - Río Piedras Campus)*; Eduardo Nicolau (University of Puerto Rico - Río Piedras Campus)

This research aims to develop a bioinspired membrane enhanced with WS2 nanosheets for operation in forward osmosis mode, thereby integrating the benefits of polymeric and semiconducting materials in the remotion of model pollutants and avoiding biofouling formation. This membrane was synthesized by modifying the polysulfone (PSF) polymer backbone with the zwitterionic group diethyl amine N-oxide (DEAO) to increase the hydrophilic properties of the final material. Subsequently, tungsten disulfide (WS2) nanosheets were introduced into the polymeric solution to integrate its photosensitive characteristics into the resulting membrane. The synthesized membranes were meticulously analyzed utilizing scanning electron microscopy (SEM), surface Z-potential measurements, and water contact angle (CA) assessments.

Furthermore, the efficacy of the fabricated membranes was evaluated based on water permeance and their capability to reject model dyes rhodamine B (RhB) and sulforhodamine (SRh). The results indicate that the photodegradation of the RhB is approximately 50% after 5 hours of light exposure, and the adsorption of SRh is above 80% after the same 5 hours without light. The dye remotion performances of these membranes are nearly 30% and 75% higher than those of the control membranes, respectively. In addition, the membranes with WS2 nanosheets embedded were able to inhibit the proliferation of Escherichia coli on their surface. The WS2 nanosheets are probable to generate reactive oxygen species (ROS), which expedites the degradation process in the presence of light to remove the RhB and prevent biofouling formation. Meanwhile, the SRh is removed from the water by the electrostatic interaction with the membrane's surface. The overall approach of this work may pave the way for the fabrication of membranes with multiple functionalities.

Membranes for Rapid Protein Analysis

Merlin Bruening (Notre Dame)*; Junyan Yang (University of Notre Dame); Weikai Cao (University of Notre Dame)

Functionalized membranes can rapidly purify, digest, and desalt proteins, thus simplifying sample preparation prior to protein analysis. Convective mass transport and short diffusion distances make these processes especially rapid. As an example, membranes can serve as a platform for rapid analyses of monoclonal antibodies during protein manufacturing. After capture of the monoclonal antibodies, binding of a labelled secondary antibody yields a fluorescent signal that is proportional to the quantity of primary antibody captured. Membranes in 96-well plates enable convenient assays in 5 minutes when using a vacuum manifold to pull solutions through the membrane. When embedded in spin tubes, membranes provide a particularly convenient platform for protein digestion and desalting. Stacking of a protease-containing membrane and a desalting membrane allows sequential digestion and desalting during flow through the membranes to rapidly produce peptides for protein characterization. We are working on coupling this technique to mass spectrometry for rapid analyses, including determination of glycosylation profiles that are a critical quality attribute for monoclonal antibodies. Overall, functionalized membranes can play a range of roles in simplifying and shortening sample preparation for analysis.

Bringing MTR's Polaris™ membrane CO2 capture process to commercial scale

Richard Baker (Membrane Technology & Research Inc.); Phillip Behm (Membrane Technology & Research Inc.); Alicia Breen (Membrane Technology & Research Inc.); Brice Freeman (Membrane Technology & Research Inc.); Pingjiao Hao (Membrane Technology & Research Inc.); Jenny He (Membrane Technology & Research Inc.); Thomas Hofmann (Membrane Technology & Research Inc.); Ivy Huang (Membrane Technology & Research Inc.); Jay Kniep (Membrane Technology & Research Inc.); Tim Merkel (Membrane Technology & Research Inc.); Witopo Salim (Membrane Technology & Research Inc.)*; Ray McKaskle (Trimeric Corporation); Dana Pierik (Sargent & Lundy); Kevin Lauzze (Sargent & Lundy); William Morris (Wyoming Integrated Test Center)

With the support of the U.S. Department of Energy (DOE), Membrane Technology and Research, Inc. (MTR) has been developing a post-combustion carbon capture system based on our Polaris™ membrane. Prior research, development, and field testing has brought this point source capture approach to a technology readiness level (TRL) of 6. This previous work includes >11,000 hours of small pilot testing at the National Carbon Capture Center (NCCC) at the 1 tonne CO2/day (TPD) scale, and later at the 20 TPD scale. More recently, small pilot testing of the Gen-2 Polaris membrane and advanced planar stack modules was completed at the Technology Centre Mongstad (TCM) in Norway.

The success of this previous work led to the Large Pilot project (DE-FE0031587) with the end goal of building and operating a 150 TPD small commercial-sized system treating 10 MWe of flue gas at the Wyoming Integrated Test Center (WITC) located adjacent to Dry Fork Station (DFS) coal-fired power plant in Gillette, WY. The Large Pilot system includes elements that are part of MTR's full-scale commercial process design such as the flue gas pretreatment and conditioning; two stages of membrane unit operations; and a CO2 purification unit. Just as important, the system also features MTR's containerized membrane skids and planar membrane modules, both in their final commercial form factor. The 150 TPD system is currently in operation producing high-purity, liquid CO2 and is the largest membrane capture system in the world. Operation of this system has moved the MTR membrane-based capture technology for power plants to TRL-7.

This presentation will describe the design, construction, and operation of the Large Pilot. Preliminary results from field testing will be discussed. These results will be used to refine the MTR CO2 capture process for future full-scale demonstration at DFS as well as deployment at industrial sites such as cement and steel plants.

Fibril contact angle: a new metric to predict wetting resistance for membrane distillation

Shounak Joshi (University of Southern California)*; Allyson McGaughey (University of New Mexico); Amy Childress (University of Southern California)

Membrane distillation (MD) persists as a competitive process for recovering water and extracting volatile species from highly concentrated streams. MD relies on a hydrophobic membrane to support a vapor gap between a warmer, liquid feed stream and a cooler distillate. If membrane pore(s) become completely wetted (i.e., pore breakthrough occurs), the feed and distillate streams mix, resulting in process failure. Thus, understanding membrane wetting and wetting resistance are critical.

Wetting resistance in porous, hydrophobic membranes is typically indicated by either the intrinsic or measured contact angle. The intrinsic contact angle represents an ideal, smooth surface. The measured contact angle represents a Cassie-Baxter wetting state, where liquid contacts the membrane surface and trapped air. During MD, however, membranes typically operate in a Wenzel wetting state, where liquid is at or within pore entrances. To elucidate wetting resistance, determining the contact angle at pore entrances is key. However, direct measurement of microscopic contact angles at pore entrances has significant experimental challenges. Instead, we introduce the fibril contact angle θ_f, calculated using measured fibril roughness and the Wenzel model. θ_f values are significantly lower than measured contact angles but higher than intrinsic contact angles. We investigate the accuracy of θ_f in characterizing wetting resistance by comparing LEP model results to wetting behavior in situ using electrochemical impedance spectroscopy. We define the fibril LEP as the transmembrane pressure (TMP) at which the wetting front progresses independently of TMP increase (i.e., the point at which TMP overcomes pore wetting resistance). Our results show that θ_f can provide better prediction of wetting front progression compared to measured and intrinsic contact angles. Moreover, to impart wetting resistance, fibril roughness may be as or more important than membrane roughness.

Ion Selective In-plane Transport in Stacked Layers of Graphene Oxide

JUMI DEKA (VANDERBILT UNIVERSITY)*; PAVAN CHATURVEDI (VANDERBILT UNIVERSITY); PIRAN KIDAMBI (VANDERBILT UNIVERSITY)

Ion transport in reconstructed layered materials, such as graphene oxide (GO), molybdenum disulfide (MoS_2), boron nitride (BN), layered hydroxides, and MXenes, presents a fundamental challenge in both scientific understanding and technological applications. A key aspect is deciphering the mechanisms governing in-plane and through-plane ion mobility while evaluating how nanochannel dimensions influence ion selectivity. Addressing these challenges is crucial for optimizing these materials for energy storage, desalination, molecular sieving, and sensing technologies.

This study focuses on the in-plane conductivity of stacked GO layers, systematically investigating key factors affecting ion transport. Specifically, we explore (i) the role of variations in channel dimension in ion filtration and conductivity, showing enhanced conductance for protons as compared to other hydrated cations. (ii) The influence of ion hydration states on mobility and selectivity within confined nanochannels revealing preferential transport of monovalent cations (K^+/Li^+) over divalent cations (Mg^{2+}). (iii) The impact of dynamic variations in channel dimensions due to physical confinement or environmental conditions. The study examines the effect of physical confinement and hydration on (GO) films within flexible polydimethylsiloxane (PDMS) and rigid epoxy resins. PDMS allows GO swelling upon water and electrolyte exposure, expanding interlayer spacing and disrupting structural confinement, thereby altering ion transport. In contrast, epoxy resins restrict swelling, maintaining consistent interlayer spacing and reducing proton conductance by up to 100 orders of magnitude. By understanding these factors, we aim to establish a framework for controlling ion transport in layered materials, paving the way for their application in next-generation filtration and energy systems.

Solution-friction approximation as a robust theoretical framework for modeling low-salt-rejection reverse osmosis

Rayan Alghanayem (Vanderbilt University)*; Rui Chen (Vanderbilt University); Weifan Liu (Vanderbilt University); Shihong Lin (Vanderbilt University)

Low-salt-rejection reverse osmosis (LSRRO) has been proposed to achieve minimal liquid discharge. It has been shown that LSRRO can treat highly concentrated solutions with moderate pressure. Understanding the salt transport in LSRRO membranes is critical to optimizing and further improving LSRRO technology. In this study, the solution-friction approximation model (SF-A) is employed to describe salt transport in LSRRO. In contrast to the solution-diffusion model that has been widely used for membrane desalination processes, the SF-A model quantifies membrane performance using three parameters: water permeability, salt permeability factor, and charge factor. Thus, observed salt permeability becomes a function of salt concentration and charge factor. We tested oxidized membranes at various feed salt concentrations (0.1 to 3.7 M NaCl) and various permeate fluxes (10 to 70 LMH). We fitted the extracted parameters from the SF-A model to our experimental flux-rejection data. In conclusion, this research demonstrates that the SF-A model effectively predicts salt transport in LSRRO membranes, offering a robust theoretical framework for accurately modeling and optimizing LSRRO systems.

Thin Film Composite Ion Exchange Membranes by Electrospray

Yasmin Teixeira Bimbatti (UCONN)*; Deniz Ipekçi (UCONN); Xingyu Wang (UCONN); Edward Wazer (UCONN); Inara Oliveira (University of Virginia); Anna Harris (University of

Virginia); Hazel Gerber (Georgia Institute of Technology); Baikun Li (UCONN); Geoffrey Geise (University of Virginia); Paul Kohl (Georgia Institute of Technology); Jeffrey R. McCutcheon (UCONN)

Producing thin film composite polymeric membranes via electrospray has been demonstrated to be a promising approach to overcome several drawbacks present in traditional solution and extrusion casting processes. The drawbacks include the large consumption of expensive selective materials and/or solvents, as well as the production of thick films. Thick films result in significant electrical resistance and, consequently, high energy consumption in electrochemical systems, directly impacting the operating costs of such systems and limiting the application space. Employing electrospray as a membrane manufacturing technique allows the formation of thin film composite ion-exchange membranes with tailorable properties and composition, without requiring excessive use of materials. Moreover, this approach has the potential to be extended to commercial materials such as Nafion or nonfluorinated alternatives that may emerge as more sustainable solutions. This research focuses on the fabrication and characterization of printed thin-film composite Nafion membranes using electrospray, in which a commercial ionomer dispersion is electrosprayed to produce thin, low-resistance, and selective cation-exchange membranes. Here, we investigate the ability of electrospray-printed Nafion ionomer to form ultrathin film composite membranes for diverse applications, integrating its unique properties and the electrospray printing approach. This technique enables precise tuning of membrane properties by adjusting the thickness of the selective layer, and/or by changing the substrate, which significantly influences membrane performance. For example, a 5 μm-thick selective Nafion layer was formed on the surface of a porous substrate and thermally annealed, resulting in membranes with a permselectivity of 0.89 compared to N115 (127 μm, permselectivity: 0.94), a significant performance achievement given the membrane's reduced thickness.

High-Performance Hemofiltration via Molecular Sieving and Ultra-Low Friction in Carbon Nanotube Membranes

Peifu Cheng (Vanderbilt University)*; Nicholas Ferrell (The Ohio State University Wexner Medical Center); Carl Öberg (Lund University); Steven Buchsbaum (Lawrence Livermore National Laboratory); Melinda Jue (Lawrence Livermore National Laboratory); Sei Park (Lawrence Livermore National Laboratory); Dan Wang (The Ohio State University Wexner Medical Center); Shuvo Roy (University of California, San Francisco); Francesco Fornasiero (Lawrence Livermore National Laboratory); William Fissell (Vanderbilt University Medical Center); Piran Kidambi (Vanderbilt University)

Kidney disease is a serious global health problem and can lead to kidney function loss and kidney failure, where renal replacement therapy like hemofiltration/hemodialysis is needed. An ideal hemofiltration/hemodialysis membrane should allow for high hydraulic permeability and maximize the removal of middle molecules while maintaining albumin losses in a clinically acceptable range. However, most commercial membranes are currently made of polymers with unevenly distributed polydisperse, tortuous and rough pores, which offer limited hydraulic permeability and imprecise sieving performances between undesirable middle molecules (~15–60 kDa) and albumin (~66 kDa). Here, we demonstrate scalable bottom-up fabrication of wafer-scale carbon nanotube (CNT) membranes with highly aligned, low-friction, straight-channels/capillaries and narrow pore-diameter distributions (sub-5 nm), to overcome inherent separation challenges in hemofiltration/hemodialysis membranes. Using FITC-Ficoll 70 and albumin in phosphate-buffer saline (PBS) as well as in bovine blood plasma, we show that CNT membranes can allow for significantly higher hydraulic permeability (more than an order of magnitude when normalized to pore-area) than commercial high-flux hemofiltration/hemodialysis membranes (HF400), as well as greatly enhance removal of middle molecules while maintaining comparable albumin retention. Our work presents potential to aid miniaturization of dialyzer sizes for progress towards wearable, portable, and implantable hemofiltration/hemodialysis devices.

Direct membrane osmometry for measuring osmotic pressure of concentrated multi-electrolyte solutions

Jeffrey McCutcheon (University of Connecticut)*; Mi Zhang (University of Connecticut); Edward Wazer (University of Connecticut); Wajeha Tauqira (University of Connecticut); George Bollas (University of Connecticut)

Accurate measurement of osmotic pressures in brines is a critical need when designing emergent osmotic membrane processes such as high-pressure reverse osmosis, osmotically assisted reverse osmosis, and low salt rejection reverse osmosis. We describe a new direct membrane osmometer that measures the osmotic pressure of concentrated brine solutions containing both single and mixed electrolytes. Unlike conventional osmometry methods like vapor pressure osmometry and freezing point osmometry, this method enables the direct measurement of a hydrostatic pressure that a saline solution exerts in a fixed volume environment when placed opposite a pure water solution across a semi-permeable membrane. Our process enables quick measurements (less than 5 minutes to equilibrium) with an exceptionally high capacity to measure osmotic pressure (2800 psi). Model brine solutions with single electrolytes (e.g. NaCl) show a greater than 95% agreement with literature predictions. Multi-electrolyte solutions show more interesting results as combinations of electrolytes fail to follow the traditional rules of colligative properties of solutions. Combined electrolytes in certain cases show synergistic effects, where osmotic coefficients are higher than anticipated. Knowing these deviations from ideal solution behavior is important when designing osmotic processes as the pumps and system components must be chosen to operate at pressures above the actual osmotic pressure of these feed solution at an anticipated recovery.

Tuning membrane properties and process conditions for preserving functional properties of biological macromolecules

Lidietta GIORNO (National Research Council of Italy - Institute on Membrane Technology (CNR-ITM))*

The use of artificial membranes for the hosting and/or processing of biological macromolecules, such as proteins, is of great interest in various fields, including bioseparation, biocatalysis, and biosensing. In all applications, the structural, physico-chemical, and morphological properties of the membrane play a decisive role. In the case of an enzyme-loaded membrane, for example, the biocatalytic activity, selectivity, and stability of the biohybrid system depend on the rigidity/flexibility of the heterogenized enzyme and the interfacial energy between the biohybrid construct and the reagent. In addition, the transport rate and the residence time of the reagent through the porous matrix of the membrane have a major influence on the overall performance. The presentation will illustrate case studies where enzyme activity is significantly different when the same enzyme is bound to the same membrane material (using the same methodology) and the membrane is in environments with different wettability, affecting not only the conformation of the macromolecule but also the interaction of the reagent with the active site. In addition, by tuning the path length (membrane thickness), the residence time can be modulated while the flow rate through the membrane remains constant. It is shown that the performance is better when the enzyme is bound to a spherical topography. Examples of the performance of biofunctionalized membranes for biosensing applications are discussed.

Acknowledgments
Authors acknowledge the project "Tech4You - Technologies for Climate Change Adaptation and Quality of Life Improvement" financed by the European Union – Next Generation EU

Protein-Enabled Size-Selective Defect-Sealing of Atomically Thin 2D Membranes for Dialysis and Nanoscale Separations

Peifu Cheng (Vanderbilt University)*; Nicholas Ferrell (The Ohio State University Wexner Medical Center); Saban Hus (Oak Ridge National Laboratory); Nicole Moehring (Vanderbilt University); Matthew Coupin (The University of Texas at Austin); Jamie Warner (The University of Texas at Austin); An-Ping Li (Oak Ridge National Laboratory); William Fissell (Vanderbilt University Medical Center); Piran Kidambi (Vanderbilt University)

Atomically thin 2D materials present potential for advancing membrane separations via a combination of high selectivity (from molecular sieving) and high permeance (due to atomic thinness). However, the creation of a high density of precise nanopores (narrow-size-distribution) over large areas in 2D materials remains challenging and non-selective leakage from nanopore heterogeneity adversely impacts performance. Here, we demonstrate protein-enabled size-selective defect sealing (PDS) for atomically thin graphene membranes over centimeter scale areas by leveraging the size and reactivity of permeating proteins to preferentially seal larger nanopores (≥4 nm), while preserving a significant amount of smaller nanopores (via steric hindrance). Our defect-sealed nanoporous atomically thin membranes (NATMs) show stability up to ~35 days during size-selective diffusive separations with a model dialysis biomolecule fluorescein isothiocyanate (FITC)-Ficoll 70 in phosphate buffer saline (PBS) solution as well as outperform state-of-the-art commercially available dialysis membranes (molecular-weight-cut-off ~3.5–5 kDa and ~8–10 kDa) with significantly higher permeance for smaller solutes KCl (~0.66 nm) ~5.1–6×10-5 ms-1 and vitamin B12 (B12, ~1.5 nm) ~2.8–4×10-6 ms-1 compared to small protein lysozyme (Lz, ~4 nm) ~4–6.4×10-8 m s-1, thereby allowing unprecedented selectivity for B12/Lz ~70 and KCl/Lz ~1280. Our work introduces proteins as nanoscale tools for size-selective defect sealing in atomically thin membranes to overcome persistent issues and advance separations for dialysis, protein desalting, small molecule separations/purification and other bioprocesses.

A membrane process for the purification of extracellular vesicles from milk whey and their application as drug delivery systems

Sara Giancaterino (Università di Bologna); Caterina Piunti (Università di Padova); Riccardo Onesti (Università di Bologna); Elisa Cimetta (Università di Padova); Cristiana Boi (Università di Bologna)*

Extracellular vesicles (EVs) are membrane-enclosed nanoparticles that are naturally released by cells and play a key role in intercellular communication, making them attractive candidates for the diagnosis of various diseases, for drug delivery, and their efficacy in regenerative medicine has been demonstrated. Regardless of their origin, EVs are currently isolated by sucrose gradient centrifugation, immunoaffinity chromatography or differential ultracentrifugation (UC), the latter being the benchmark. These techniques allow for high recovery yields, but high costs and low scalability hinder their production on an industrial scale.

This work presents an economical and scalable membrane filtration process for the isolation of EVs from whey derived from the production of Parmesan cheese. Several experiments were performed to determine the optimal operating conditions, paying particular attention to the membrane type and the filtration modes (i.e., concentration and diafiltration). The isolated vesicles were analyzed and characterized by physical and biochemical methods. An optimized membrane ultrafiltration (UF) process can lead to a reduction of impurities up to 99.8%, allowing the continuous production of a purer stream compared to UC (although less concentrated), containing up to 4·1011 EVs/mL with a mean hydrodynamic size around 180 nm and a surface charge of -11.8 mV. This process was used to develop a drug delivery system (DDS) for neuroblastoma (NB) therapy using curcumin as a model drug. The efficacy of the process was evaluated by cell viability and internalization studies on SK-N-AS NB cells.

Curcumin-loaded EVs showed higher cellular uptake and greater inhibition of SK-N-AS cell proliferation compared to free curcumin, with a lower IC50 (16.34 μM vs. 31.14 μM). This scalable and efficient method for the production of Cur-loaded EVs from bovine milk whey shows potential for therapeutic applications in cancer treatment.

Scalable Bottom-Up Synthesis of Nanoporous Hexagonal Boron Nitride (h-BN) for Large-Area Atomically Thin Ceramic Membranes

Andrew Naclerio (Vanderbilt University); Peifu Cheng (Vanderbilt University)*; Saban Hus (Oak Ridge National Laboratory); Trey Diulius (National Institute of Standards and Technology); Marti Checa (Oak Ridge National Laboratory); Ivan Vlassiouk (Oak Ridge National Laboratory); William Fissell (Vanderbilt University Medical Center); Matthew Coupin (The University of Texas at Austin); Jamie Warner (The University of Texas at Austin); Liam Collins (Oak Ridge National Laboratory); Andrei Kolmakov (National Institute of Standards and Technology); An-Ping Li (Oak Ridge National Laboratory); Piran Kidambi (Vanderbilt University)

Nanopores embedded within monolayer hexagonal boron nitride (h-BN) offer possibilities of creating atomically thin ceramic membranes with unique combinations of high permeance (atomic thinness), high selectivity (via molecular sieving), increased thermal stability and superior chemical resistance. However, fabricating size-selective nanopores in monolayer h-BN via scalable top-down processes remains non-trivial due to its chemical inertness and characterizing nanopore size distribution over large area remains extremely challenging. Here, we demonstrate a facile and scalable approach of exploiting the chemical vapor deposition (CVD) process temperature to enable direct incorporation of sub-nanometer/nanoscale pores into the monolayer h-BN lattice, in combination with manufacturing compatible polymer casting to fabricate centimeter-scale nanoporous-atomically-thin-ceramic-membranes. We leverage diffusive-transport of analytes including size-selective Ficoll sieving to characterize sub-nanometer and nanoscale defects that manifest as pores in centimeter-scale h-BN membranes, overcoming previous limitations in large-area characterization of nanoscale defects in h-BN. Our approach opens a new frontier to advance atomically thin membranes to 2D ceramic materials, such as h-BN via facile and direct formation of nanopores, for size-selective separations.

Methods for Measuring Experimental Energy Barriers to Transport in Salt-Rejecting Membranes

Mohammad Allouzi (University of Colorado Boulder)*

Exploring the fundamental factors that govern water and solute transport through salt-rejecting membranes is essential for improving their separation efficiency and selectivity. Measuring experimental energy barriers has become an increasingly popular tool for gaining a mechanistic understanding of transport. This transport framework, based on permeability measurements at different temperatures, provides insights into molecular-level phenomena within the membrane and at the solution-membrane interfaces. Although the energy barrier approach has yielded valuable insights, it can sometimes produce inconsistent and inconclusive results. This study aims to provide a reliable interpretation of energy barriers for ion transport across loose and tight salt-rejecting membranes. Using the existing database of energy barriers in transmembrane permeation, we highlight data variability across studies and the limited ability to distinguish trends between different solutes and membrane, emphasizing the need for more reliable quantification methods. Next, we explore the role of temperature cycling in energy barriers determination, demonstrating that membranes equilibrated at the highest operating temperature and tested in a descending temperature order produce more stable and reliable trends compared to those exposed to increasing temperatures. Following thermal stabilization, evaluating energy barrier results at a similar concentration polarization modulus is essential for a reliable interpretation of trends across different solutes and membranes. Applying this protocol in two independent laboratories, we obtain energy barriers for ion permeation that align with the performance of loose and tight salt-rejecting membranes. Overall, this study underscores universal recommendations to improve the experimental implementation of energy barrier measurements in salt-rejecting membranes.

Block Copolymer Adsorptive Membranes for PFAS Removal

Annabelle Li (University of Notre Dame)*; Siwei Gu (University of Notre Dame); Kyle Doudrick (University of Notre Dame); William Phillip (University of Notre Dame)

Adsorptive membranes are a promising technology for the fast, efficient capture of per- and polyfluoroalkyl substances (PFAS) from drinking water. PFAS can flow through the ligand-containing pores of the membrane and adsorb more quickly than in a traditional packed bed. However, membranes have low binding capacities due to their limited pore wall chemistries. To address these challenges, the surface segregation and vapor induced phase separation (SVIPS) process has been used to fabricate highly porous polysulfone/polystyrene-block-poly(acrylic acid) (Psf/PS-PAA) composite membranes with chemically-tunable PAA-lined pores. For instance, the PAA brushes can be reacted to attach polyethylenimine (PEI) ligands that provide additional binding sites. The Psf/PS-PEI membrane can achieve equilibrium binding capacities of 496 mg PFOS and 302 mg PFOA per gram of membrane, which exceeds that of conventional adsorbents. Further investigation of this system involved fabricating membranes using PS-PAA copolymers with different molecular weights and molar ratios of polystyrene to polyacrylic acid. Small changes to the PS-PAA composition caused large differences in the membrane structure and quantity of available binding sites. Future work to establish the relationship between copolymer composition and membrane structure will allow for the rational design of membranes optimized for PFAS capture.

High-Performance Water-Based Membranes for Efficient and Stable Carbon Dioxide Separation

Kian Lopez (University of Colorado Boulder)*; Max Saffer-Meng (University of Colorado Boulder); Mohammad Allouzi (University of Colorado Boulder); Yukai Tomsovic (University of Colorado Boulder); Anthony Straub (University of Colorado Boulder)

Efficient carbon dioxide separation is necessary for carbon capture, biomethane purification, and spaceflight life support. However, existing membrane-based technologies often suffer from low permeability, insufficient selectivity, and instability in humid conditions, limiting their effectiveness in practical applications. Conventional sorption-based materials degrade over time or require high energy input, making large-scale deployment challenging. To address these limitations, we have developed a gas separation membrane that uses a pressure-stable liquid water layer within hydrophilic pores. This design takes advantage of the high solubility of CO_2 in water, enabling selective transport while restricting the passage of other gases. Membranes with a thin (<200 nm) water layer achieve CO_2 permeabilities ten times higher than the best-performing membranes in the literature while maintaining a CO_2:N_2 selectivity over 40:1. Gas transport through the membrane is governed by solubility-driven diffusion, improving CO_2 separation efficiency without the need for complex chemical processes. The membranes remain stable under continuous operation for at least one week without performance degradation. Additionally, they can withstand pressures exceeding 70 bar without structural failure, demonstrating their suitability for high-pressure industrial applications. The water-based separation mechanism also tolerates high humidity feed streams, overcoming a common limitation of conventional gas separation materials that degrade or lose performance in moist environments. This membrane technology provides a scalable, energy-efficient approach to carbon capture by leveraging a water-based transport mechanism. Its combination of high permeability, selectivity, and operational stability makes it a strong candidate for use in carbon removal systems across various industries.

Employing MOF-808-Functionalized NF-Membrane for Selective Extraction of Selenium from Produced Water

Sweta Modak (The University of Alabama)*; Milad Esfahani (The University of Alabama)

Produced water from oil extraction, with high salinity (~35000 ppm) and elevated selenium (Se) concentrations (500 ppm-300 ppt), poses challenges for reuse. Se, carrying bioaccumulation risks, can be detrimental when ingested above 0.05 ppm, affecting aquatic life, wildlife, and humans. Nanofiltration (NF) membranes have been explored for treating produced water, showing effective removal of Se oxyanions exceeding 93%. However, NF membranes also reject pollutants, necessitating an additional step for safe recycling. Se-selective adsorptive membranes facilitate adsorption during filtration, eliminating the need for further separation. This study presents MOF-adsorptive membranes that leverage the advanced properties of metal-organic frameworks (MOFs) to selectively extract Se from high salinity and oily waste streams, such as produced water. Given the satisfactory removal performance of MOF-808 (~75 mg/g selenite), it was selected for chemical grafting onto TFC membranes to develop MOF-membranes. This study examined the effects of various components (salts, oils, and organic matter) on the MOF-membrane performance. The bare TFC membrane demonstrated rejection rates of ~95%, ~20%, and ~12% for 2000 ppm Na_2SO_4, NaCl, and 35000 ppm NaCl, respectively. The MOF-808-membrane exhibited ~92%, ~8%, and ~5% rejection at the same concentrations. While MOF-808-membranes showed slightly lower salt rejections, the water permeability of MOF-808-membrane increased from ~3 LMH/bar to ~10.5 LMH/bar compared to bare TFC membrane. The bare TFC membrane rejected ~86% and ~83% of 3 ppm selenite and selenate, respectively, while the MOF-808 membrane removed ~53% and ~74% of these ions at the same concentration. Although the MOF-membrane exhibited a lower rejection of selenite, it showed ~25% adsorption of selenite within 60 min at room temperature. The MOF-membrane exhibited excellent regenerability, maintaining consistent adsorption efficiency even after five adsorption cycles.

Hydrogen Sulfide and Hydrocarbon Separation Properties of CTA Hollow Fiber Membranes for Natural Gas Sweetening

Dan Miller (SLB)*; Atsushi Morisato (SLB); Ed Mahley (SLB)

Natural gas processing is one of the largest industrial markets for gas separation membranes. Virtually all natural gas must be treated to remove carbon dioxide (CO_2) and hydrogen sulfide (H_2S) before entering a pipeline, as these “acid gases” can corrode the pipeline. While there has been significant effort to understand separation of binary mixtures of CO_2 and methane (CH_4), separation of H_2S from complex hydrocarbon mixtures is not frequently reported due to the complexity of multi-gas measurements and the safety measures required for work with H_2S. In this presentation, H_2S separation from realistic natural gas hydrocarbon mixtures with commercial asymmetric cellulose triacetate (CTA) hollow fiber membranes will be discussed. Like CO_2, H_2S is a plasticizing gas, so its presence in the feed gas mixture can affect the membrane permeability and selectivity. The gas mixture used for this study was 49% CH_4, 15% C_2H_6, 10% C_3H_8, 5% CO_2, and 21% H_2S. The mixed-gas permeation tests were performed at three temperatures, 15 °C, 25 °C, and 35 °C, and four feed pressures, 15 bar, 29 bar, 39 bar, and 46 bar. This presentation will explore the relationships among feed gas pressure, temperature, and composition to understand how highly condensable plasticizing gases impact CTA hollow fiber performance in natural gas sweetening. In some feed gas pressure regimes, including the lowest feed pressures tested, the selectivity of the membrane for acid gases over hydrocarbons was essentially unaffected by plasticization. However, the selectivity of the membrane for acid gases over the three hydrocarbons in the feed gas mixture was compromised as feed pressure was increased due to the greater mobility of the CTA polymer chains under plasticizing conditions. This presentation will describe how feed pressure and temperature can be leveraged to optimize acid gas separation from hydrocarbons in natural gas sweetening.

Lack of Chemical Stability of Metal Organic Frameworks (MOFs) in Mixed Matrix Membranes for Aqueous Separation Applications

Matthew Davenport (The University of Texas at Austin)*; Kevin Reimund (The University of Texas at Austin); Matthew Dombchik (The University of Texas at Austin); Benny Freeman (The University of Texas at Austin)

Metal Organic Frameworks (MOFs) have been considered for selective separations based on their well-defined, tunable pore size and facile incorporation of different chemical functionalities. They exhibit strong size-sieving properties for monovalent aqueous ion separations (e.g., Li+/Na+), and they have already found commercial applications for gas separations, such as sorbents for food packaging and electronic gas delivery for sensitive applications. However, commercialization of MOFs for aqueous separation applications remains elusive, potentially due to instability in water. This study evaluates desorption of linker molecules and the corresponding degradation of MOFs in water—both when immersed directly in liquid water and when integrated into a mixed-matrix membrane. Four MOFs identified as water-stable in the literature with relevance to ion separations are evaluated: UiO-66, UiO-66-(COOH)2, UiO-66-NO2, and UiO-66-NH2. None were found to be, in fact, be stable. Kinetics of linker desorption were studied, with power-law kinetics observed. Standard MOF characterization methods, including XRD, SEM imaging, BET surface area, and XPS were performed before and after soaking in aqueous solutions. Large defects were observed in mixed-matrix membranes following immersion in water. The effect of external salt concentration and speciation is considered. In summary, no MOF that we studied was stable in water.

Engineering polymeric membranes for high-performance air dehumidification

Faheem Akhtar (LUMS)*

The water vapor-selective membranes present a unique opportunity to greatly reduce the power requirements for moisture removal by avoiding phase change and have thus been ranked as a top alternative to traditional HVAC systems. Various systems incorporating hygroscopic salts introduced may offer important alternatives. Salt solutions embedded in graphene aerogel fibers are highly promising for high flux because of the high-water uptake capacity. The main challenge for their realization is incorporating a high density of salts and desorption of the adsorbed moisture, especially in arid climates. Another way of incorporating hygroscopic salts, desired membrane performance can be achieved by intrinsically designing a polymer system. In this presentation, I will make a critical comparison of these two routes, discuss the origin of the intrinsic modification in polymer, and share our recent success in boosting the performance of membrane-based systems for air dehumidification in the air-conditioning industry.

I will discuss the advantages and disadvantages of rational design of polymeric materials to engineer membranes, as example technologies developed in our lab. For example, sulfonated poly(ether-ether-ketones) (SPEEKs) with modular degree of sulfonation. However, the answer might not be so definitive. I will share our recent results that show that the in silico and experimental investigations at a molecular level enabled the development of systems exhibiting outstanding water uptake capacity of more than 300%, fast sorption uptake, and high transport rate. Membrane performance analyses demonstrated remarkably high-water vapor permeability and selectivity; therefore, the membranes developed herein showed potential for application in water vapor control and dehumidification processes in enclosed or confined spaces. Contrary to common assumptions, the correlation between the degree of sulfonation and dehumidification performance showed a plateau after maximum curvature.

Continuous Production of API Nanocrystals by a Porous Hollow Fiber Membrane-based Novel Anti-solvent Crystallizer

Yufeng Song (New Jersey Institute of Technology); Lixin Feng (New Jersey Institute of Technology); Wei Zhong (New Jersey Institute of Technology); Siddharth Tripathi (New Jersey Institute of Technology); Padmini Ananthoji (Amgen); Piero Armenante (New Jersey Institute of Technology); Kamalesh Sirkar (New Jersey Institute of Technology)*

In porous hollow fiber membrane (HFM) based anti-solvent crystallization (AsCr) technique developed by us and investigated by others, the anti-solvent is injected from HFM bore into the active pharmaceutical ingredient (API) containing solution flowing in the shell side. In this design, the shell-side feed solution flows along the HFM. We have studied a totally different membrane module configuration and design, where the API-containing solution flow path length is ~ 3 cm reducing considerably the residence time (tres) which strongly influences crystal growth. It allowed us to obtain API nanocrystals which strongly enhance the bioavailability for a majority of APIs which are hydrophobic. We investigated AsCr of griseofulvin (GF) from its acetone solution using anti-solvent water coming through HFM pores. The range of tres was 5-30 s; the volume flow rate ratio (FRR) of acetone/water ranged between 0.35 to 1.66; griseofulvin concentration was varied between saturation (3.73 g/100 g acetone) to 0.4 g GF in 100 g of solution having 50 wt% acetone in the total solution. Low tres yielded nanocrystals <100 nm. The average API crystal size (Davg) remained around 230 nm in a three-hour-long experiment. At a fixed mass FRR of 1.34± 0.1, Davg increased linearly from 48 to 280 nm as tres increased from 5 to 28 s. For two HFM modules in series with the nanocrystal suspension from one module fed to the next module independently fed with anti-solvent water, GF crystallization yield as high as 98% was achieved. AsCr of L-glutamic acid in water with anti-solvent acetone in the same device illustrated continuous nanocrystal production and similar behavior vis-à-vis variation of tres and FRR. This HFM module based AsCr technique can continuously produce nanocrystals of APIs.

Methanol separation from poly(lactide) depolymerization products

Suer Kurklu Kocaoglu (The University of Texas at Austin)*; Sabah Najjar (The University of Texas at Austin); Carson Womack (The University of Texas at Austin); Louise Kuehster (The University of Texas at Austin); Hamidreza Mahdavi (Monash University); Nirvana Almada (The University of Texas at Austin); Laila Halim (Monash University); Michael Batten (CSIRO); Leonie van't Hag (Monash University); Zongli Xie (CSIRO); Zachariah A. Page (The University of Texas at Austin); Matthew R. Hill (Monash University); Nathaniel A. Lynd (The University of Texas at Austin); Benny D. Freeman (The University of Texas at Austin)

Plastic management and recycling are essential processes for achieving a sustainable future. Poly(lactide) (PLA) is considered a sustainable polymer because it is derived from renewable sources such as corn starch and is biodegradable. However, while PLA production continues to increase, its biodegradation requires high temperatures and a humid environment for complete breakdown, which can take several months.

As a result, many studies focus on the recycling and depolymerization of PLA. One of the most prevalent recycling processes is methanolysis, which offers the advantage of producing methyl lactate. However, product separation remains a challenge. In our study, we systematically investigated PLA methanolysis using the catalyst 1,8-Diazabicyclo(5.4.0)undec-7-ene (DBU) and explored the degradation reaction mechanism. Although all reactions ultimately led to complete depolymerization into methyl lactate over longer reaction times, methyl lactate conversion reached $96.1 \pm 2.4\%$ within 25 minutes under optimal conditions. Under these conditions, the depolymerization product consisted of approximately 50 wt% methanol-methyl lactate mixture.

To obtain methyl lactate for potential use as a green solvent or a reactant for lactide production, we tested various commercial nanofiltration (NF) and reverse osmosis (RO) membranes. Using a dead-end filtration setup, we characterized the water and methanol permeability of the membranes and further evaluated their separation performance.

Investigating the Energetics of Condensable Gas Transport in Triptycene-Poly(arylene ether) Membranes

Erin Sturd (Massachusetts Institute of Technology)*; Zachary Smith (Massachusetts Institute of Technology)

Preliminary data has shown that microporous polymers, including poly(arylene) ethers, can be promising materials for condensable gas separations, such as hydrofluorocarbon refrigerant reclamation[1] and carbon dioxide purification from industrial off-gas[2-4]. However, industrial use of these materials remains challenging due to several issues including diffusion-driven separations[1,5] and plasticization[6], when condensable gases dissolve into the polymer resulting in accelerated segmental motions and decreased size-sieving. In this talk, sorption and permeability results are reported for a series of triptycene-poly(arylene ethers) (PAEs) for CH4, CO2, CH2F2, and CF4. Using the dual-mode model and pure-gas permeation experiments, the solubility and diffusivity effects were deconvoluted to determine the mechanism of sorption and sorption-selectivity. Observed improvements in sorption selectivity for the non-fluorinated and partially fluorinated polymers were found to result primarily from increased Langmuir sorption selectivity. Preliminary analysis of energetics indicates that fluorination does not appear to affect bulk penetrant-polymer mixing as strongly as fractional free volume. Finally, CO2/CH4 sorption selectivities are reported above unity, indicating a method to increase sorption selectivity for condensable gas separations.

1. Gutiérrez-Hernández, S. V. et al. J. Mem. Sci. 2023, 675, 121532.
2. Yi, S. et al. J. Mater. Chem. A 2015, 3 (45), 22794–22806.
3. Guo, S. et al. Angew Chem Int Ed Engl. 2024, 63(8), e202315611.
4. Genduso, G. et al. Journal of Membrane Science 2019, 584, 100–109.
5.Harders, A. N. et al. Ind. Eng. Chem. Res. 2023, 62 (9), 4054–4063.
6. Rodriguez, K.M. et al. Chem. Soc. Rev. 2024, 53, 2435-2529.

Performance evaluation of commercial RO membranes in treating produced water with effective membrane cleaning approach

Md Hasib Al Mahbub (Texas Tech University)*; Mahdi Malmali (Texas Tech University)

Produced water (PW) is a byproduct of oil and gas extraction, rich in dissolved salts and organic substances, necessitating treatment due to water scarcity and environmental regulations. This study examines the tertiary treatment of PW using five commercially available RO membranes and proposes an effective method for mitigating scaling and fouling to recover membrane flux. Five different flat-sheet membranes were purchased from DuPont FilmTech and Hydranautics in this work for PW treatment. Membranes were characterized for water contact angle, FTIR , SEM, and EDS analysis. PW was collected from a mid-well oil and gas field with TDS upto 30000 ppm. Filtration tests of the PW with the RO membranes were run targeting 60% feed recovery. Of the five different membranes, SW30HRLE was found to show the highest flux and salt rejection. Major salt ions in the permeate were analyzed with ICP-MS, and the results showed 99.2 to 99.8% rejection of Na, Ca, Mg, Sr, etc. A further thorough study was performed cleaning the SW30HRLE membrane fouled by PW with two different in-place cleaning strategies. The membrane cleaning includes a high pH (pH = 11) and a low pH (pH = 2) cleaning agent. These membrane cleaning techniques were employed between the consecutive PW filtration tests of a 3-cycle experiment. The cleaning experiments show that base and acid cleaning recovered 79.63% and 92.12% of the initial flux. EDS analysis of both the low pH and high pH treated membrane showed that effective cleaning of the membrane surface was achieved with the high pH cleaning compared to the acid cleaning. This study demonstrates a comparison of different commercial membranes' performance in PW treatment and an effective cleaning strategy of the membrane, which allows high water productivity and provides potential research direction for industrial applications in PW treatment.

Industrial potential of nonwoven membrane chromatography adsorbers for the purification of biologics

Cristiana Boi (Università di Bologna)*; Jinxin Fan (North Carolina State University); Joseph Lavoie (North Carolina State University); Luca Masciolini (North Carolina State University); Riccardo Onesti (Università di Bologna)

Nonwoven membranes with conformal UV grafted layers around each fiber, and modified with ion exchange and multimodal ligands, exhibit high dynamic binding capacities for biologics at residence times of less than one minute. This enables their use for membrane chromatography processes with productivities at least an order of magnitude higher than conventional resin-based column chromatography. Our group has used these membranes for the isolation and purification of different biopharmaceutical targets, from monoclonal antibodies and other proteins, to adeno-associated viral vectors, DNA and RNA.

These novel membranes enable all-membrane processes to replace column chromatography for product capture, purification and final polishing. The advantages of this approach are discussed in the context of a first test of an all-nonwoven membrane process for the purification of a monoclonal antibody from mammalian cell culture. Most of this work has been done with membrane volumes of 2 mL or less, with the membranes packed into 1 cm diameter columns for even flow distribution. This paper will also describe current efforts in our laboratory involving the scale-up of the membrane grafting and functionalization process, and the development of a membrane module that features excellent chromatographic performance and can be applied in parallel to accommodate the needs of large-scale processes. Both objectives need to be met for the industrial application of this technology.

In particular, the membrane modules being developed overcome the drawbacks of currently available commercial membrane devices used in biopharmaceutical manufacturing for filtration, clarification and impurity removal, namely: low binding capacities, large dead volumes, and poor fluid distribution resulting in wide breakthrough and elution curves than are obtained with resin chromatography.

Selective recovery of rare earth elements by membrane-based solvent extraction process

Victoria Bessah (North Carolina A&T State University); Shailesh Dangwal (Oak Ridge National Laboratory); Ajaratu Cole (North Carolina A&T State University); Syed Islam (Oak Ridge National Laboratory); Shamsuuddin Ilias (North Carolina A&T State University); Ramesh Bhave (Oak Ridge National Laboratory); Md Arif Khan (North Carolina A&T State University)*

The sustainable recovery of rare earth elements (REEs) is indispensable for modern technologies, from renewable energy systems to advanced electronics to electric vehicles.

However, their conventional extraction and separation are energy-intensive, environmentally damaging, and inefficient in separating chemically similar REEs. Membrane-based solvent extraction (MSX) offers a promising alternative by combining the precision of solvent extraction with the efficiency of membrane technology. However, studies on MSX processes lacked comprehensive optimization and integration strategies for industrial scalability and sustainability as the interplay between mass transfer, diffusion, and reaction kinetics within the membrane system is not fully understood, limiting the ability to build a robust predictive model for process optimization and scale-up. Addressing these challenges is essential for ensuring a sustainable and secure supply of these critical materials for the future, supporting global efforts toward cleaner energy and resilient technological development.

This study seeks to demonstrate the efficiency and sustainability of membrane-based solvent extraction (MSX) for rare earth element recovery by optimizing process parameters, modeling mass transfer and reaction kinetics, and integrating with downstream recovery systems for industrial scalability. The experimental solvent extraction results are validated by a physicochemical model for the MSX process based on selective reaction and diffusion through the membrane pores, accurately predicting REE recovery and separation. Key parameters were optimized using experimental data, achieving excellent agreement and robust scalability confirmed by Monte Carlo simulations. The findings highlight MSX as a sustainable and efficient solution for REE recovery and highlights MSX as an efficient and sustainable solution for REE recovery.

Transport of Ionic Chemotherapy Drugs in Charged Biosponge Block Copolymers

Hee Jeung Oh (Pennsylvania State University)*

Cancer is the leading cause of death in most developed nations. Despite the effort to develop targeted and personalized chemotherapy drugs, the dosing of drugs is limited by toxic side effects. More than 90% of the injected chemotherapy drug, regardless of injected amount, is not trapped by a target organ, goes to the heart, and circulates throughout the body. In the context of reducing chemotherapy toxicities, we have designed biosponge adsorbers that can selectively bind to target ionic chemotherapy drugs. In this work, we aim to understand the governing transport mechanism of ionic chemotherapy drugs in charged nanostructured block copolymer membranes with different ion-exchange capacities (IECs) and equilibrium water swelling. Sulfonated stryrenic pentablock copolymers are used as a model polymer, and doxorubicin (dox) is used as a model chemotherapy drug. We performed a cycle of drug sorption/desorption experiments to quantify the amount of free-dox and bound-dox in the sulfonated polymer membranes. Using a swine model, our initial design enables the capture of 69 % of the administered drug without any adverse effects. Upon this work, we aim to understand the design principles of charged biosponge block copolymers for different ionic chemotherapy drugs. Designing highly-selective biosponge polymer adsorbers can effectively remove untrapped chemotherapy drugs in the body, and thus, can help patients fight cancer.

Solid-state NMR methods for quantifying composition and crosslinking in polyamide-based thin film composite reverse osmosis membranes

C. M. Stafford (NIST)*

Establishing clear relationships between the chemical structure, water / ion transport, and performance in the selective layer of thin film composite (TFC) reverse osmosis (RO) membranes is important in water filtration design, yet these relationships are difficult to establish. In this talk I will show solid-state NMR results on the polyamide-based selective layer in four trimesoyl chloride (TMC)/isophthaloyl chloride (IPC)/meta-phenylene diamine (MPD)-based TFC membranes that were fabricated at the pilot plant at Dupont Water Solutions. In these films the crosslink density was intentionally reduced by replacing trifunctional TMC monomers with their difunctional IPC analogue. In these films we directly measured the crosslink density and composition from our 13C solid-state NMR spectra. Furthermore, we demonstrate that (1) chemical purity is measured directly from the NMR spectra, which is important for industrial RO membranes, and (2) low field magic angle spinning (MAS) NMR is useful for 13C quantitation, which is used to calculate populations of the TMC, IPC, and MPD monomer units. Our results show that both crosslink density and polarity are important design criteria in RO membranes, since they affect the local diffusion coefficient, as observed from quasi-elastic neutron scattering experiments. We also extended our 13C CPMAS methodology to bulk polymerized polyamides with varied chemical structure, demonstrating the robustness of the methodology. Results of diffusion NMR on these bulk polymerized polyamides will be given as well, which will demonstrate the importance of length- and time-scale in reporting measured diffusion coefficients.

Electroconductive Ultrafiltration Membranes Enhanced with Nickel-Doped Laser-Induced Graphene for Efficient Removal of Emerging Pollutant

Swatantra Singh (IIT BOMBAY)*; Ashish Kumar (IIT Bombay)

New generation membrane technology has globally fascinated the scientific community as an advance alternative of conventional treatment methods for water purification due to their compact treatment process and efficient pollutant removal. However, the high-pressure requirement and fouling tendency of high-pressure driven membrane such as nanofiltration (NF) and reverse osmosis (RO) is still a challenge considering the processing cost and lifespan of membrane. Electroconductive ultrafiltration (UF) membranes immobilized with nano-catalyst present a groundbreaking solution for efficient pollutant removal via electrochemical advanced oxidation process (EAOP). In the present work, we have fabricated the Nickel doped laser induced graphene (LIG) based electroconductive UF membranes utilizing a CO2 infrared laser. These membranes capitalize on the exceptional properties of Ni-LIG, including high catalytical properties, electrical conductivity, surface area, crucial for efficient electrochemical reactions. The fabrication of these Ni-LIG electroconductive composite membranes on the polymer surface was confirmed through various spectroscopic and electron microscopic techniques. Moreover, their electrochemical activity was checked and demonstrated through various electrochemical characterization tools such as cyclic voltammetry and electrochemical impedance spectroscopy. Furthermore, these Ni-LIG electroconductive composite membranes has demonstrated their effectiveness in removal of emerging contaminants in flow through filtration mode. Additionally, the incorporation of Ni nanomaterial exhibited improved performance resulting in 20 - 30 % more removal of emerging contaminants as compared to its counterpart, thanks to its electro-catalytical properties. Nevertheless, the integration of LIG into ultrafiltration membranes enables simultaneous filtration and pollutant degradation, offering a promising avenue for sustainable water treatment.

Fabrication of Asymmetric Selective Bilayer Membranes with Tight Pores at the Interface and Micro-pores within the Matrix using Dual-Layer Slot Die Coating on a Roll-to-Roll

Minwoo Jung (Georgia Institute of Technology)*; David Lu (University of Kentucky); Eesh Kulshrestha (University of Kentucky); Isabel Escobar (University of Kentucky); Tequila Harris (Georgia Institute of Technology)

Bilayer membranes are extensively studied for water filtration applications for their to enhance the selectivity of membranes or add different functional materials for each layer to increase the effectiveness of the filtration performance. Numerous studies have concentrated on the pore and filtration properties of individual layers, but few have examined the interface between two layers and the pores that this interface creates. When a bilayer membrane is fabricated using two different polymers, tight pores at the interface can naturally occur, depending on the material and the process.

In this study, an alternative approach for fabricating tight pores at the interface in the bilayer membrane using two just different polymers is developed. Here, dual-layer slot die coating is used to fabricate two distinct, A and B, polymer membrane fluid streams simultaneously, which are subsequently cured using phase inversion. By slowing the phase inversion rate of the bottom layer polymeric membrane A due to the top polymeric membrane B, a selective tight-pore layer is created at the interface. The sample was delaminated to observe the pores at the interface surface. The bottom surface of the top layer polymeric membrane B exhibited an average pore size of 0.4 μm, the largest pore being 4 μm from the sample observed. The top surface of the bottom layer of polymeric membrane A exhibited very tight pores that were not visible with 90k magnification with SEM. However, the low permeation data at a low pressure of 60 psi validates that the bilayer membrane was permeable), hence the pore exists at the nanometer range. The results from this work show that pore tightness can be introduced using a single-pass manufacturing process, and the resulting pores can be exploited as an in-situ selective layer.

Rational design of oriented mixed-matrix metal-organic framework membranes for molecular separations

Shuvo Datta (KAUST)*

Traditional chemical separation methods for valuable commodities are highly energy-intensive, consuming approximately 10–15% of global energy. A promising alternative for achieving energy-efficient, safe, and sustainable separation processes is the use of metal-organic framework (MOF)-based mixed-matrix membranes (MMMs). These advanced membranes synergistically combine the selective adsorption and molecular sieving capabilities of MOFs with the mechanical stability and processability of polymers. In gas separation applications, the development of energy-efficient MMMs requires high-aspect-ratio MOF nanosheets with fully exposed pores/channels to facilitate selective gas diffusion and enhance MOF-polymer interactions. However, significant challenges remain, including the development of MOFs capable of withstanding harsh gas conditions and the precise tuning of their morphology into oriented nanosheets that align within the polymer matrix. To overcome these challenges, we recently developed a novel methodology for the rational design and fabrication of highly efficient, oriented mixed-matrix MOF (MMMOF) membranes. By integrating three interlocked design criteria, these membranes demonstrated exceptional performance in separating H2S and CO2 from natural gas under practical operating conditions.[1] This innovative approach represents a breakthrough in MMM technology, offering transformative potential for a variety of energy-intensive industrial separations, including the highly challenging ethylene/ethane separation.[2]

Co-treatment of municipal wastewater and crop harvest waste by an anaerobic membrane bioreactor for providing water, nutrients, and energy to controlled environment agriculture

Abayomi Alayande (Clemson University)*; Weiming Qi (Clemson University); Sudeep Popat (Clemson University); David Ladner (Clemson University); Gary Amy (Clemson University)

An anaerobic membrane bioreactor (AnMBR) can effectively treat municipal wastewater and provide high-quality irrigation water containing plant nutrients to controlled environment agriculture (CEA). However, a key challenge is the low organic carbon content, which results in a low organic loading rate (OLR), potentially limiting chemical oxygen demand (COD) removal and methane production. An AnMBR typically exhibits low nitrogen and phosphorus removal, a positive attribute for crop cultivation. Two strategies to enhance OLR include (i) reducing hydraulic retention time (HRT) and (ii) co-treating organic-rich waste streams. This study investigates an AnMBR-ultraviolet (UV) system designed for hydroponic CEA applications, focusing on improving in situ nutrient recovery and methane generation. Specifically, we assess whether incorporating ground/screened or pre-fermented crop harvest wastes from hydroponic controlled-environment agriculture (CEA) enhances COD removal, methane production, and effluent nutrient enrichment (e.g., nitrogen and phosphorus). Three scenarios are evaluated: (i) treatment of synthetic municipal wastewater alone, (ii) co-treatment with ground/screened crop harvest waste, and (iii) co-treatment with pre-fermented crop harvest waste. The system employs a tubular silicon carbide ceramic membrane (200 nm pore size, 0.072 m^2 surface area) operated at a fixed HRT of 0.75 days and a constant flux of 8.4 LMH, with intermittent backwashing (5 s every 30 min). We hypothesize that integrating crop harvest waste will significantly enhance nutrient recovery and methane yield, offering a sustainable approach to resource recovery in municipal wastewater and hydroponic CEA systems.

Elucidating flowrate effects in anion-exchange membrane processes for continuous production of adenovirus therapeutics

Claire Velikonja (McMaster University)*; William Pihlainen-Bleecker (McMaster University); Cindy Shu (McMaster University); Adrian Foell (McMaster University); Landon Steenbakkers (McMaster University); Brandon Corbett (Sartorius); Chris McCready (Sartorius); Prashant Mhaskar (McMaster University); David Latulippe (McMaster University)

Purification is a major bottleneck in virus therapeutic manufacturing, accounting for over 70% of operational costs. Continuous chromatography (CC) has reduced costs in protein manufacturing by using multiple chromatography units in complex configurations; however, its application to viruses remains underexplored. Due to their larger size (~10x that of proteins), viruses require chromatography materials without mass-transfer limitations, with anion-exchange membranes emerging as a viable option. A key challenge in virus CC adoption is understanding how optimal CC operating variables change with different virus-membrane combinations.

Adenovirus is one of the most used virus vectors, used in 50% of ongoing gene therapy clinical trials. While trends have been identified in flowrate (high flow rates result in improper virus binding, while low flowrates encourage irreversible binding of the virus), these trends have been reported for lentiviruses, which are enveloped, in comparison to non-enveloped adenoviruses.

Therefore, there is a need to evaluate key operating variables, such as flow rate, for non-enveloped viruses like adenovirus to enable virus CC design.

This study investigates flowrate effects on adenovirus using Sartobind Q anion-exchange membranes. A high and low flow rate (5 and 30 membrane volumes/min) were tested with five chromatography cycles at each condition, with host-cell DNA, proteins, and genomic and infectious titer measurements taken from the fractions. UV spikes were seen during the 1 M NaOH cleaning phase, indicating the presence of irreversibly bound material. The % DNA removal and infectious titer recovery did not vary with respect to flow rate. Further work will assess the protein removal and the genomic vs. infectious titer ratio and characterize the species eluted in the 1 M NaOH phase. This study highlights the distinct behavior of adenoviruses in contrast to lentiviruses, underscoring the need for virus-specific CC process optimization.

Effect of Widespread Stacking Disorder on Single Crystal Covalent Organic Framework (COFs) Membrane Performance

Yun Kyung Shin (Pennsylvannia State University); Beatrice Bartolomei (Northwestern University); Caleb Johannesmeyer (University of Wyoming); Will Dichtel (Norwestern University); Adri van Duin (Pennsylvania State University); Katie Li-Oakey (University of Wyoming) *

Ideally, separation materials possess long range order, are tailorable to specific applications, and simultaneously achieve high flux and high selectivity. Previous work by our collaborative team demonstrated discrepancy between experimental observation and force field modeling. Defects in polycrystalline COF layers have been attributed to the discrepancy. While single crystal COF has been reported, there has been a lack of direct experimental evidence on interlayer stacking of single crystal COF until recently. Additionally, due to the brittle nature of single crystal COFs, there is no current report on membranes made from single crystals. In this talk, we report COF membranes fabricated using single crystal TAPB-DMPDA COF [TAPB:1,3,5-tris(4-aminophenyl) benzene; DMPDA:2,5-dimethoxyterephthalaldehyde]. Combined with a reactive force field (ReaxFF) model, the effect of interplanar stacking on the single crystal COF membrane performance is characterized and corroborated with experimental filtration results. To minimize the time and length scale mismatch between the theoretical predication using the atomistic ReaxFF modeling and experimental results, a common liquid, such as water, is used to normalize the diffusion rate calculated by the ReaxFF model and flux results from filtration tests, respectively. The direct comparison between ReaxFF model calculation and filtration results reveals how the stacking disorder affects COF membrane porosity and selectivity, providing insights for advancing COF design and synthesis, and bridging the gap between their theoretical properties and practical applications.

Characterization of polymer thin films by atomic force microscopy

Olivia McIntee (National Renewable Energy Laboratory)*; Nurshaun Sreedhar (National Renewable Energy Laboratory); Brian Welch (Israel Institute of Technology); Victor Bright (University of Colorado Boulder); Abhishek Roy (National Renewable Energy Laboratory);

Mou Paul (National Renewable Energy Laboratory); Alan Greenberg (University of Colorado Boulder)

Polymer membranes for water and energy applications must withstand high pressure, temperature, and/or extreme pH during operation and cleaning procedures. Recent research has highlighted the impacts of membrane morphology on mechanical behavior and performance. Thus, morphology is a key parameter when developing and evaluating novel membrane materials.

Atomic force microscopy (AFM) offers unique advantages for nano-scale characterization of thin film membrane interfaces and morphologies. In this study, the mechanical behavior of crosslinked polyamide thin films was evaluated using AFM in force spectroscopy mode under simulated operating and/or cleaning conditions. Reverse osmosis thin films were fabricated using (1) interfacial polymerization (IP), a liquid-phase reaction where film thickness is self-limiting; (2) a new method, molecular layer deposition (MLD), a gas-phase deposition technique that can create conformal thin films with precisely controlled chemical composition and thickness.

The results show that in the dry state the modulus of the MLD film (3.4±0.91 GPa) was nominally higher than the IP films (2.1±0.89 GPa). Given that the films operate in the hydrated state, companion testing was conducted in water, and data indicated that both MLD and IP film stiffness decreased significantly. After chlorine exposure the MLD film maintained 61% of its hydrated modulus whereas the IP films only retained 40-54% of their hydrated moduli, suggesting that the stiffer MLD films may be more durable than the IP films. AFM roughness scans indicate the MLD films are smooth and do not have the rough peak-and-valley topography of IP films, which may be a factor in the differences in mechanical behavior between the MLD and IP films. Additional membrane chemistries will be explored using this methodology to further understand membrane structure-property relationships. These include polyimide IP films, bipolar membranes, and polybenzimidazole (PBI) membranes.

Turn
into options
opportunities
MilliPORE
SiGMa
SAFC® Millipore®

the human energy company®
Chevron
the brightest minds help
illuminate the way forward
Pushing new ideas to solve energy challenges is progress at work. At Chevron, we're applying technology in innovative ways to explore what's possible. This includes developing cutting-edge software to analyze data more effectively, as well as partnering with universities and research institutions to help pioneer new solutions. All to fuel innovation that helps us meet the world's energy demands now and for many years to come.
Learn more at Chevron.com
Chevron proudly supports
the North American Membrane Society.